FOLDABLE FLEX AND THINNED SILICON MULTICHIP PACKAGING TECHNOLOGY

**A publication by Kluwer for IMAPS
(International Microelectronic and Packaging Society)**

EMERGING TECHNOLOGY IN ADVANCED PACKAGING SERIES

Advisory Council

IMAPS and Kluwer Academic Publishers have set up an advisory board for the Emerging Technology Series of books. They not only recommend titles, but also review proposals and suggest direction. All are experienced professionals who have published many papers, articles, books and journal contributions. All have been given many honors for technical leadership in the electronic packaging Industry.

Series titles:

1. Foldable Flex and Thinned Silicon Multichip Packaging Technology

FOLDABLE FLEX AND THINNED SILICON MULTICHIP PACKAGING TECHNOLOGY

edited by

John W. Balde
IMAPS Fellow
IEEE Fellow

KLUWER ACADEMIC PUBLISHERS
Boston / Dordrecht / London

Distributors for North, Central and South America:
Kluwer Academic Publishers
101 Philip Drive
Assinippi Park
Norwell, Massachusetts 02061 USA
Telephone (781) 871-6600
Fax (781) 681-9045
E-Mail: kluwer@wkap.com

Distributors for all other countries:
Kluwer Academic Publishers Group
Post Office Box 322
3300 AH Dordrecht, THE NETHERLANDS
Telephone 31 786 576 000
Fax 31 786 576 254
E-Mail: services@wkap.nl

  Electronic Services <http://www.wkap.nl>

Library of Congress Cataloging-in-Publication Data

Foldable flex and thinned silicon multichip packaging technology / edited by John W. Balde.
     p. cm. – (Emerging technology in advanced packaging series ; 1)
  Includes bibliographical references and index.
  ISBN 0-7923-7676-5
   1.   Microelectronic packaging. 2. Multichip modules (Microelectronics) 3. Chip scale
     packaging. 4. Flexible printed circuits. I. Balde, John W. 1923- II. Series

TK7870.15.F65 2002
621.381'046 – dc21

                         2002034095

Printed on acid-free paper.

Printed in the United States of America.

Contents

Contributor List

Chapter 1
John W, (Jack) Balde
    Interconnection Decision Consulting
    Flemngton, New Jersey USA
    Phone: 1-908-788-5190
    Fax: 1-908-782-3351
    e-mail: balde@IDConsultng.com

Chapter 2
Dr Evan Davidson
    IBM (Retired)
    Hopewell, New York. USA
    Phone: 1-845-221-8437
    e-mail: Evan.Davdson@ieee.org

Chapter 3
Christine Kallmyer
    Fraunhofer Institute for Reliability and Microintegration (IZM-B)
    Berlin, Germany
    Phone: 49 30 46403 228
    Fax: 49 30 46403 161
    e-mail: kallmayr@izm.fhg.de

Chapter 4
Thomas Harder, Wolfgang Reinert
    Fraunhofer Institute for Silicon Technology
    Itzehoe, Germany
    Phone: 49 (0) 4821 / 17 - 4620
    Fax: 49 (0) 4821 / 17 - 4690
    e-mail: harder@isit.fhg.de

Chapter 5
Dr. Karlheinz Bock, Michaek Feil, Christof Landesberger
    Fraunhofer Institute for Reliability and Microintegration (IZM-M)
    Munich, Germany
    Phone: 49 (0) 89 54759-506
    Fax: 49 (0) 89 54759- 550
    e-mail: bock@izm-m.fhg.de
    michael.feil@izm-m.fhg.de
    christof.landesberger@izm-m.fhg.de

Chapter 6
Dr.Jørg-Uwe Meyer
  Dråger AG
  Lubeck, Germany
  Phone: 48-451-882-3295
  Fax: 48-451=882-72256
  e-mail: Jeorg-uwe.meyer@drager.com

Chapter 7
 Michael Warner
  Tessera Technologies
  San Jose, California USA
  Phone: 1-408-383-3609
  Fax: 1-408-894-0768
  e-mail: MWarner@Tessera.com
William Carlson
  Tessera Consultant
  San Jose, California USA
  Phone: 1-408-383-3690 or
         1-408-997-2403
  Fax: 1-408-894-0285 or 408-997-2409
  e-mail BCarlson@Tessera.com or

Chapter 8
 Georges Rochat, Philippe Clot, Jean-François Zeberli
  Valtronic SA
  Les Charbonnieres
  Switzerland
  Phone: 41-21 841 0111
  Fax: 41-21 841 0222
  e-mail: infoval@valtronic.ch

Chapter 9
 Dr.Ted Tessier
  13713 SW Tracy Place
  Tigard ,Oregon, USA
  Phone: 1-503-524-0874
  Fax: same number, handshake first
  e-mail: azTessier@aol.com

Chapter 10
Jan Vardaman,
  TechSearch International
  Austin, Texas. USA
  Phone: 1-512-372-8887
  Fax: 1-512-372-8889
  e-mail: Jan@TechSearchInc.com
Dominique Numakura
  DKN Research
  Haverhill, Massachusetts, USA
  Phone: 1-978-372-2345
  Fax: 1-978-469-0188
  e-mail: dNumakura@ATT.Global.com

Chapter 11
Dr. Rui Yang , Dr. Terry F. Hayden
  3M Microinterconnect Systems Division
  Austin, Texas USA
  Phone: 512-984-5682
  Fax: 512-984-5940
  e-mail: tmhayden@mmm.com

Chapter 12
 Dr. Leonard Schaper
   University of Arkansas
   Fayettevile, Arkansas, USA
   Phone: 1-501-575-8408
   Fax: 1-501-575-2719
   e-mail: schaper@uark.edu

Chapter 13
 Dr. Carl Zweben
   62 Arlington Road
   Devon, Pennsylvania, USA
   Phone: 1-610-688-1772
   Fax: 1-610-688-8340
   e-mail: c.h.zweben@usa.net

About IMAPS

IMAPS, the International Microelectronic and Packaging Society, is the result of a merger of two leading technical societies for the electronic packaging community. The original society, the International Society for Hybrid Electronics, invariably known as ISHM - was founded in the USA in 1967 with a principal focus on ceramic hybrid assemblies and materials.

Over the years ISHM grew to encompass chapters in all parts of the world, giving it a true international representation. Within its portfolio it included a wide range of substrate, interconnect and assembly technologies providing a broad forum for discussion and development of expertise.

In 1996 ISHM merged with IEPS, the International Electronic Packaging Society, an organization founded in 1978 with electronic systems emphasis, including activity in chip carriers, multichip modules and computer systems.

IMAPS fits these thrusts together to form a unified society dealing with the totality of electronic system design, packaging and materials. It organizes major conferences and symposia in the US, Europe and Japan, and many workshops on the specialized subjects of electronic packaging. Local chapters organize many activities and the society has various publications including a bi-monthly bulletin and a Journal.

It has led in the development of many new technologies and, with this series, intends to lead in the development of a range of advanced technologies relevant to the industry.

IMAPS 'Emerging Technology' Books

IMAPS, the International Microelectronics and Packaging Society, is sponsoring a series of books on *Emerging Technologies in Advanced Packaging*, edited by John Balde of Interconnection Decision Consulting. Unlike anthologies or compendia of published papers, most of the material in these books is not yet published, or published only in limited form.

In a sense, this is a new concept for technical books – a pre-anthology book of very new material.

The intent is to focus on the edge of new technology – discussing technology that is not yet in general usage, but shows promise to produce major change in the industry. These are not handbooks – standard material and fundamentals are to be found elsewhere, and will be referenced. Rather, this is new directions either just being introduced or newly identified as major technology to be pursued further.

Patterned after the Radiation Laboratories series of books in the late 1940's, it is reviewed and directed by an advisory board of leading packaging technologists in the world, both in the US and overseas.

The series will deal with Folded Flex Packaging; Photonics and Planar Waveguide Technology; Medical Electronics; MEMS for Electronic Circuits; RF Technology; and BioEngineering Technology.

This is the first of those books and I recommend it to you.

Peter Barnwell
President, IMAPS

Introduction

John Balde, Editor, Fellow IMAPS, Fellow IEEE
Interconnection Decision Consulting

Foldable Flex and Thinned Silicon Chip Packaging Technology

The impetus for this book came from exposure to the new technology work of the Fraunhofer companies of Germany, Fraunhofer IZM in Munich and Fraunhofer Medical in St Ingebert. Presentations of the work with thinned chips and foldable flex were made at Workshops in Zurich, Barcelona and Newport, Rhode Island, in particular, through presentations by Dr. Rolf Aschenbrenner of Fraunhofer IZM and J-Uwe Meyer of Fraunhofer Medical.

Mostly the work was being reported only at Workshops, but then Fraunhofer ISM (Silicon Technology) did a presentation at IMAPS, France on the activities of European consortium or project to develop technologies and standard operations for the silicon die thinning for flex use. A Project called Flex/Si had progressed to the reporting stage.

It was obvious that this technology was flourishing in Europe, but largely ignored in the rest of the world.

It was being used to make stacked chip packages in the so-called 2-1/2 D technology (3-D in physical format, but interconnected only through the circuits on folded flex.) It was also being used in single chip packages where the thinness of the chips and the flex substrate made packages significantly thinner than through any other means.

The two figures on the following pages from Fraunhofer IZM declare the intent of the technology and the nature of the single chip assemblies as well.

Concept for this Book

This book is organized to report on the developments in this technology, but with special additional material and emphasis.

The intent is to do more than report on present state of the art. It is intended as an advocacy book, pointing out the reasons for 3-D assemblies, the reasons for Silicon-in-a-Package multichip modules, and the commercial availability of the techniques.

It will also do more than discuss the present, it will point out the deficiencies of the constructions, the needed availability of good flex material, the use of newer flex materials such as LCP, and the implications from the use of the Integrated Mesh Power Systems to enhance the capability for future designs. Lastly it will discuss the serious problem of heat removal if multiple microprocessors are included.

Flexible Substrates

Material	Polyimid „Kapton"
Lines	Cu, Ni, Au
Tg	> 400 C
Dielectric	3 - 4
Line and Space	75 / 75 µm, typ.
Via-Diameter	200 µm
Layers.	1 … 4
CTE	27 ppm
Therm. Cond.	0,16 W/mK

„CHiP"

CHip in Polymer

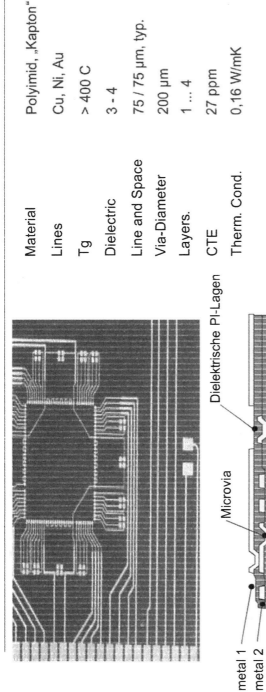

Dielektrische PI-Lagen

Microvia

Polyimid

metal 1
metal 2
metal 3
metal 4

Fraunhofer Institut
Zuverlässigkeit und
Mikrointegration

Combination of Flexible Substrates and Thin Chips

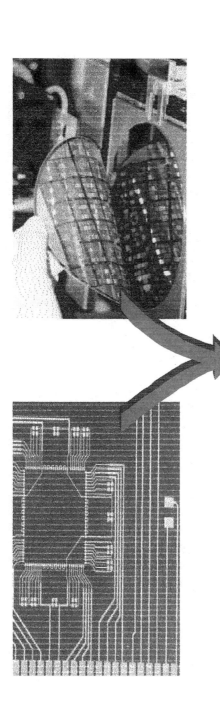

Ultra Thin Packages

Packages with Special Geometries (flexibility)

Thin 3-D Stacks

Embedding of Thin Chips in Substrates

Fraunhofer Institut
Zuverlässigkeit und
Mikrointegration

Specific Contributions

The first two chapters are exploratory and are analyses of alternative assembly constructions. The varieties of 3-D assemblies and their deficiencies, the comparison of the efficiency and electrical performance of Silicon-on-a-Chip (SOC) versus Silicon-in-a Package (SIP) technologies will be explored.

Dr. Evan Davidson of IBM has spent much time in analyzing the relative performance and cost / yield of these two constructions, even for SIP MCM technologies on ceramic or MCM-L constructions using conventional chips. He will discuss the potential for foldable flex and thinned silicon chip MCMs to provide high interconnectivity and low profile and low cost.

Exposition of the Work in Germany

The next three chapters present the state of the technology as practiced by the two divisions of Fraunhofer IZM in Berlin and Munich. The concepts originated in Berlin, and Christine Kallmeyer will present the concepts and many of the applications already implemented in these constructions.

The work of chip thinning has been taken to a high level or process refinement at Fraunhofer Munich, and Dr. Karlheinz Bock will present that work.

These processes have led to the coordination of the work in this area through the Flex/Si project, with Dr. Thomas Harder presenting the state of the concensus on the right technologies.

The Fraunhofer Medical Work

Fraunhofer Medical in St. Ingbert Germany has taken the technology for use in biomedical devices. In some ways the work has a little different focus. The chip thinning is accepted as a given, but their focus was on very dense compact flex circuits, down to 10 Micron line and space interconnections. Not only are their circuits unique, but their die attachment by through-via riveting is also special.

Through their techniques they had have many types of medical implant devices, including the development f an implantable artificial retina to restore sight to the blind. Folded stacked multichip assemblies so small to be implantable in the vitreous humor of the eyeball shows their skills.

Dr. Jorg-Uwe Meyer will present their technology and their applications.

Commercial Package Vendors

This foldable flex technology as developed by the Fraunhofer companies, is in the design mode, with various companies going into production with the Fraunhofer designs. They are not a manufacturing house.

But there are two packaging manufacturers who use the foldable flex concept, both with ordinary chips and thinned chips, who provide packages for customers.

The US supplier, Tessera, presents their technology in the chapter by Mike Warner, Dr. Georges Rochat presents the technologies and capabilities of Valtronic as a European source of such packages.

Critical Analysis

In a departure from the usual, there is a chapter about why this technology in not in favor in the United States. AMKOR, working with INTEL, came to the conclusion that the processes were not yet suitable for them to consider. They will explore the perceived deficiencies and the needs to be satisfied to permit future use and expansion of the application of this technology.

This chapter will identify the need for low cost, high density flex, and the need to control crosstalk and power and ground so that the assemblies are more useful than just for stacked memory chips. The question of heat removal becomes a problem when microprocessors are included instead of just memory.

Response to the Perceived Critical Issues

The next chapters deal with those issues. Jan Vardaman will deal with the multiple sources of suitable flex with emphasis mostly on polyimide. Dr. Terry Hayden of 3-M will focus on the great advantages through the use of LCP material for the flex. It is not just the much better performance and lower cost, but the superior dimensional control that reduces via pad size and increases circuit density capacity.

Increasing performance of circuits with multiple microprocessor chips requires control of power and ground and crosstalk, but with a two sided flex that can be folded. Dr. Leonard Schaper of the University of Arkansas describes the Integrated mesh Plane Power Systems that bring the power and ground connections into the double sided flex circuits. This may well be the enabling technology that permits folded flex multichip construction that permits this technology to be used for many silicon-In-a-Package applications. Though 2-1/2 D technology may never fully compete with stacked via-connected die, it can do many jobs at much lower cost.

Heat Removal

If it can compete for MCM SIP applications with high heat producing chips, it cannot do so unless heat removal methods improve. Dr. Carl Zweben will explore the possibility for materials that can enhance the heat removal from dense packages.

Overview

This book, therefore, explores the needs, reveals the state of development and production, but also points to changes in technology that can bring this technology into wider use for more complex applications. It is an advocacy book in this respect - advocacy for the use of a technology that is already mature, and advocacy for exploring ways to make it even more capable for the future.

BIBLIOGRAPHY

1. Rolf Aschenbrenner, Fraunhofer IZM, Germany, "Micro Megatronic Systems for Automotive Applicatioins " IEEE European Systems Packaging Workshop, Bardelona, Spain, Jan. 22 2001

2. Jurgen Wolf and Herb Reichl, Fraunhofer IZM , Germany. "High Density Interconnect and Wafer Level Packaging Trends and New Developments in Packaging", IMAPS High Density Interconnect & Systems Packaging Conference, Santa Clara, California, April 18, 2001

3. J-Uwe Meyer, H Beutel, T.Stiegliltz, D,Scholz, Fraunhofer IBMT, Germany, "Methods and Trends for Biomedical Engineering", The IEEE Internatiional Workshop on Chip-Package Co-Design. Zurich, Switzerland, March 14 2000

4. John Balde , J-Uwe Meyer, "Microflex Technology used for a Retinal Implant Device", IEEE Multichip Applications Workshop, Newport, Rhode Island, May 2001

5. Thomas Harder, W.Reinert, Fraunhofer ISIT, Germany, "Low Profile Flip Chip Assembly Using Ultra-thin ICs". IMAPS European Microelectronics & Packaging Conference", Strasbourg, France, May 31st, 2001

6. Vern Solberg, Tessera, U.S.A. "Stacked Multichip Packaging for the Next Generation Electtronics", the SMTA Pan Pacific Symposium, Maui, Hawaii, February 6 2002

7. F.Zeberti, F. Fernando, P.Clot, J.M.Chenmuz, Valtronic Switzerland, "Flip Chip with Stud Bump and Non-conductive Paste for CSP 3-D" IMAPS European Microelectronics & Packaging Conference", Strasbourg, France, May 31st, 2001

8.   John Balde, Interconnection Decision Consulting   U.S.A., "Packaging Technologies That Will Control the Future", Flextronics Conference, Sunnyvale, California, September 1998.

9.   Terry Hayden, 3-M Microelectronics. U.S.A. , "New Liquid Crystal Polymer (LCP) Flex Circuits Meet Demanding Reliability and End-use Application Requirements", IMAPS International Conference on Advanced Packaging Systems, Reno, Nevada, March 12, 2002.

10.   Evan Davidson, IBM U.S.A., "SoC or SoP ? A Balanced Approach" IEEE Electronics Componants & Technology Conference, Orlando, Florida, May 29-June 1st 2001

11.   Len Schaper, S. Ang, Y.L.Low, University of Arkansas, U.S.A., " Design of the Interconnect Mesh Power System (IMPS) MCM Topolology", ISHM / IEPS International Conference on Multichip Modules. Denver, Colorado. April 1994

12.   E. Jan Vardaman, TechSearch International" "Multiclient Report on Flex Circuit Manufacture, 2001,

13.   Carl H. Zweben, Advamced Materials Consultant, U.S.A. "Advances in Materials for Optoelectronic, Microelectronic and MEMS /MOEMS Packaging" IEEE Semi-Therm Conference, San Jose March 12-14, 2002

Chapter 1

3-D ASSEMBLIES OF STACKED CHIPS AND OTHER THIN PACKAGES

John W. Balde, Fellow IMAPS, Fellow IEEE
Interconnection Decision Consulting
Flemington, New Jersey USA

1.1 STACKING TECHNOLOGIES AND THE QUEST FOR HIGH DENSITY

In the quest for high density packaging of electronic circuits, there is no question that stacking the Integrated Circuit chips provides more capability for a given area of the mother board or substrate. The construction of co-planar multi chip modules, which decreases the surface area by removing package walls between chips, also improved the electrical performance by shortening the distance between circuit elements and removing added impedance problems and capacitances from the use of packaged individual chips. Planar multi chip modules provided that improvement, but vertical stacking of the chips can add to the circuit density per area.

If the chips of a module were stacked using vertical space, there would be a further increase in density. The question is on the type of construction to be used and its electrical and mechanical performance.

1.2 LEADED STACKED PACKAGES

This simplest of all the stacking technologies used generally off-the-shelf packages with either standard leads or perhaps specially ordered longer leads.

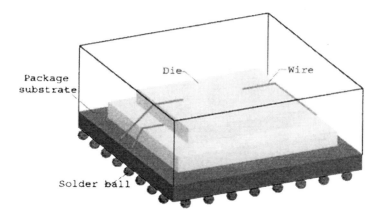

Figure 1 – Stacked Packages (*Used with permission of Steve Stankiewicz)

The lowest package is bonded to a substrate in a conventional way. The upper packages are lead bonded either onto the tops of the lower leads, or to extended conductive pads on the substrate at outer locations. But this is mostly for memory chips, in which almost all of the leads are identical and only a few need special connections.

It's Advantages:

- Standard packages, or nearly so

- Wirebonding of the die takes place at the package manufacturer - the semi-conductor house.

- Simple assembly, can be performed by most users

- It's disadvantages:

- Adds considerable height

- Requires known good die, because the assembly cannot be cheaply reworked to change any but the top-most package.

- Is limited to memory chips only

- Requires the surface area of packaged die, no reduction in footprint from that of a packaged die.

1.3 STACKED BARE DIE

The next development, shown in figure 2, was to stack bare die and connect each to the substrate by wirebonding. But simple wirebonding either requires each successive chip to be smaller than the one below, or required some means of providing clearance from the top of one chip to the bottom of the next to provide room for the wirebonding

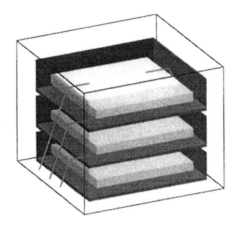

Figures 2 – Stacked Chips
(*)

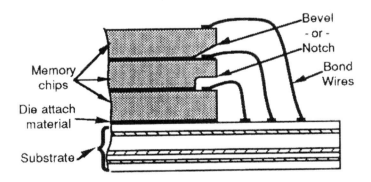

Figure 3 – Stacked Undercut Chip (*)

This called for spacers, or milling the bottom of the second or higher chips to provide the clearance for the wirebond leads, as proposed by David Tuckerman of N-Chip. Die stacking decreased the vertical height, and reduced the area somewhat because the individual die was not previously packaged.

If wirebonding was used for the assembly there were advantages and disadvantages:

Advantages:

- Standard die could be used

- Wirebonding technology could be used by the user, and did not require special operations by the die manufacturer.

- There was a reduction in height over the stacked package approach.

- The total assembly could be encapsulated into a rugged package by the user.

- Disadvantages:

- This again was only suitable for stacked memory chips - the wirebonding problem was generally too difficult for mixtures of logic and memory chips.

- It definitely needs known good die - these assemblies could not be re-worked. Testing is only on the completed assembly.

- The height was increased over the stacked height of the die alone because of the use of wirebonding.

1.4 STACKED BARE DIE USING VIA INTERCONNECTIONS

An interesting variation of the stacked bare die constructions has been proposed by Tomita and others from the ASET project at the Tsukuba Research center in Japan. Each silicon die has etched vias through the silicon substrate, and connection is made to the circuit board through these vias. The dies are connected by solder bonding of the bottom of the vias to the board.

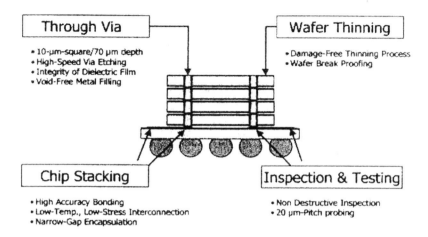

Figure 4 – Stacked Chips With Vias

Subsequent superimposed die are connected to the lower die through these via connections, bonded to the lower dies on top of the vias pads of the lower die.

This really changes the die availability and the assembly operations, providing such assemblies only from die manufacturers.

The advantages and disadvantages are more sharply drawn

Advantages:

- Reduced height over that of the wirebonded assemblies.

- Reduced inductance by the elimination of the wirebonding

- Particularly rugged and protected from mechanical damage, even before encapsulation.

Disadvantages:

- Certainly needs known good die

- Needs special die all from one manufacturer

- Limited to memory only - with identical pads

- Takes space on the lower die for the bonding pads to connect the vias from the upper die

- Takes the assembly out of the hands of the system manufacturer and moves it to the die manufacturer, with all that means for single source and limited availability.

1.5 DIE STACKING WITH INTERPOSERS

Many stacking constructions have been proposed and implemented using die bonded to thin substrates, and the substrates are then stacked, generally using via connections in the substrate. One of the first was the Stacking Technology of AT&T Bell Labs, by Segelken and Teneketges.

A more recent approach to this construction was by Piennemaa and others from Tampere University, using two chips per interposer layer.

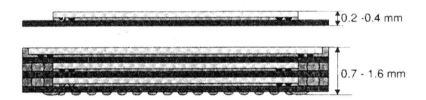

Figure 5 – Stacking With Interposers

Advantages:

- Die from different manufacturers can be used - it is the interposer that needs a standard footprint for the connection to the next layer

- Assembly is standard enough that OEMs can do it

- Underfilling is possible between layers, enhancing the reliability

Disadvantages

- Needs Known Good Die

- Adds some height because of the combined thickness of the interposers

- Generally requires memory or similar chips because of the interconnection problem

1.6 THE MONOLITHIC MOLDED BLOCK APPROACH

Stacked bare die can be interconnected by edge or peripheral connections and molded into a stacked assembly. There are three technologies:

The Irvine sensor approach used special die with all I/O brought to the ends of each die for connection at the very edge. These special die are stacked and the tiny edges of the I/O leads are electrically connected with special metallization to external connecting circuits.

This is tricky technology, requiring many different metals to make the connections. The industry analysis often said that the necessary materials used over half the metal elements in the periodic tables.

The resultant monolithic block could then be attached to a substrate, as a sort of giant thick chip.

Advantages:

- A block assembly handled like a single chip for the bonding to the substrate

- Absolutely minimal height for the stacked assembly

Disadvantages:

- Limited to memory chips, with perhaps one microprocessor chip on top.

- Definitely available from only one manufacturer, giving the control of the assembly over to that manufacturer.

- Extremely expensive and suitable mostly for special purpose military projects.

1.7 THE 3-D TECHNOLOGY OF CHRISTIAN VAL

In this technology TAB technology is used for the die bonding of each die. But the TAB frames are left connected and the frames with the die are then stacked and molded into a stacked assembly.

The stacked, potted, assembly is then sheared midway in the leads of the TAB assemblies. The resultant plastic block had the edges of the TAB leads showing on the surface of the sheared or machined face.

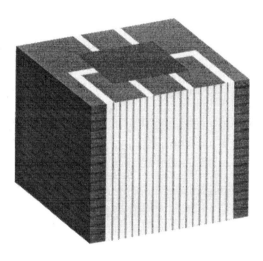

Figure 6 – 3-D+ Molded Assembly
(*)

The edge of the leads are much thicker than that of the Irvine Sensors blocks, and do not require exotic metals to make contact.

These edge contacts connected by plating of the edge surface with copper to make the connection to the leads.

The plating connects all the edges of the circuits and around the edge to the bottom of the block.   Etching away the unwanted copper leaves leads from the edges of the TAB leads of each die down the block and around the edge to provide bottom contacts,

The result is a rugged, low cost molded block that can be handled as a component.

Advantages:

- Rugged, solid block with bottom BGA contacts

- Minimal height, more than bare die but less than the stacked assembliesusing interposers

- Low cost, because the technology is well developed - the molding and plating is simple.

Disadvantages:

- Is really locked to one supplier. Cannot be performed at the OEM manufacturer.

- Generally limited to all memory, though one microprocessor chip can be on top.

- Generally limited to die that can be available in TAB bonded assemblies generally from only one manufacturer.

But both of these constructions have the same problem. The assembly has to be turned over to some third party - the OEM Manufacturer cannot do it.

1.8 FRAME WIREBONDING ASSEMBLIES

There was one possibility for OEM Manufacturers, a process called Frame Wirebonding. A variation of the 3D-technology approach was proposed by John Walker of Northern Teleom. Instead of using TAB molded die in the TAB frames for the stacking, he proposed using TAB-type frames NOT BONDED TO THE DIE. The bare TAB frames would be bonded to die using wire bonding on the TAB type lead frame to the die. This avoids the major disadvantage of the 3D approach - the need for TAB bonded dies from one manufacturer. Using the TAB wirebonding approach, die from many manufacturers could be used.

In each case the die must be pre-tested, because there is no way to change anything once the assembly is made.

1.9 NEED FOR ANOTHER APPROACH

All of these technologies are primarily for memory chips and do not constitute a replacement for multichip modules with integral passives. System-in-a-Package is possible with co-planar multichip flex modules. What is needed is a technology that might permit the inclusion of passives and microprocesors into a stacked assembly.

1.10 FOLDED FLEX ASSEMBLIES OFFER NEW OPTIONS

The concept of placing die on a long flex circuit, testing the assembly, and then folding the flex strip with either accordion folds or simpler rolled folds originated, so far as I can tell, with R.F."Jiggs" Giguere of Bell Laboratories, Holmdel in the early 1970 time period.

Werner Engelmaier of Bell Laboratories, Whippany had just explored the capability of flex circuits to withstand tight folds - 1/8 inch or so edge diameter of the fold. Previous flex work had always assumed more gradual folds - in fact the flex circuits used to connect print heads always used a 1" diameter rolling curve.

Engelmaier proved that small radius folds were reliable for up to as many as 10 folds, more than adequate for the Operations used in manufacture of assemblies that would remain in a final folded position in use.

With that information, Giguere designed assemblies for Telecom modems using the folded assembly concept. The parts could be assembled on the strip, tested, parts changed if necessary, all before the assembly was folded.

**FLEXIBLE
CIRCUIT**

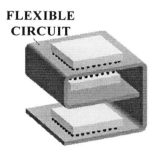

Figure 7 – A Folded Flex
Construction *

That concept remained generally unused till Fraunhofer IZM began to design using that folded flex arrangement. Their innovation was to use thinned silicon chips to make a particularly thin assembly.

Their chapter in this book will demonstrate their design and tell of their experience with that concept.

Fraunhofer Medical also used folded flex assemblies, most notably in their retinal implant and other medical applications. Their chapter of this book will show all the uses they have made of the concept, and their special Microflex film technology that produces particularly high density circuitry.

The flex circuit can be single sided or doublesided flex. The first figure shows that double sided flex is needed for some die attachment, but it is possible with proper folding techniques to use single sided flex and still mount four chips.

This technology, pioneered by Fraunhofer IZM, has produced interest throughout Europe, and a consortium has been set up to explore and document the procedures for using thin flex and thinned silicon.

Fraunhofer ISIT is the Project leader for this task force, and Thomas Harder will present the work at Fraunhofer IZM in connection with that project.

1.11 COMMERCIAL CONSTRUCTIONS

The Tessera and Valtronic folded flex constructions often use only single sided flex, with proper folding to keep the attachment of the die on one side of the flex. This simpler construction can be used for up to four die, as shown in the diagram

1.12 FOLDED FLEX CONSTRUCTIONS

These constructions can be used for stacked memory chips, but they also can accommodate microprocessor chips and other devices. They can occupy a niche more like multichip modules which can combine digital and analog chips, and other circuit elements, including passives. This puts them in the category of System-In-a-Package.

Figure 8 – A Four Chip Assembly (*)

That opens up a major capability for stacked assemblies - the use of System-in-a-Package can provide the high performance possible with multichip assemblies, that are difficult to achieve with single chip, so called System-on-Chip designs.

Let us list some of the advantages and disadvantages:

Advantages:

- Does not need pre-tested die

- Can be lower vertical height than stacked packages or stacked chips of normal thickness

- Can be assembled by the OEM - no exotic constructions that require delegating the package to a third party.

- Can be much lower cost that MCM-C or MCM-D constructions of multichip modules.

- Have multiple sources of the basic flex circuits.

Disadvantages:

- Only memory and perhaps one microprocessor chip can be placed on flex circuits of readily available density. It takes particularly high-density flex to connect disparate devices.

- Double sided flex is not readily available - sources of supply must be ascertained (The chapter by Vardaman and Numakura will explore that)

- There may be problems of crosstalk control and power managements with only double sided flex to work with - but the Integrated Mesh Power System (IMPS) of the University of Arkansas solves that problem - see the chapter by Dr. Len Schaper.

- Use for more than memory chips requires heat removal, either by interposed metal paddles or by new materials of good thermal properties. That will be explored in the Chapter by Dr. Carl Zweben.

1.13 COMMERCIAL AVAILABILITY

Both single sided flex folded assemblies and double sided flex assemblies are already commercially available through Tessera and Valtronic, and information is in their chapters of this book.

1.14 CONCLUSIONS

That this technology is already feasible and used in present products is apparent. Its potential for the future depends on the electrical performance and the capabilities of the assemblies.

1.15 BIBLIOGRAPHY

1.   Harry Goldstein, SPECTRUM Senior Associate Editor, "Packages Go   Vertical", IEEE SPECTRUM Magazine, August 2001, pp 46-51.

2.   D. B. Tuckerman et al, N-Chip, "Laminated Memory: A New 3-Dimensional Packaging Technology for MCMs", Proceedings IEEE Multichip Module Conference (MCMC) Santa Cruz, 1994, pp 58-63.

3.   C. Val and M.Leroy, Thomson CSF,"The 3-D interconnection-Applications for multichip Modules - Vertical, MCM-V",ISHM Symposium for Microelectronics, Orlando Florida, October 1991.

14

4. John Walker, Northern Telecom, Canada, "Flip Wirebonding: A strong Candidate for the KISS Approach", ISHM Thin film Multichip Modules III Workshop, Ogunquit Maine, June 17, 1992.

5. T.G.Grau, R.R.Shively, L.J.Lau, J.M.Segelken, and K.L.Tai, AT&T Bell Laboratories, "Ultra-Dense: An MCM Based 3-D Digital Signal Processor," Nepcon Conference, Anaheim, Feb 1991.

6. Seppo Pienimaa, J. Valten, R. Heikkita and E Ristolainen, Tampere University, Finland,"Stacked Thin Dice Packaging", IEEE Electronics and Technology Conference, Orlando Florida May 29, 2001.

7. Y. Tomita et al, ASET Technology Research Department, Tsukuba Research Center, Japan "Advanced Packaging Technologies on 3-D Stacked LSI utilizing the Micro Interconnection and the Layered Microthin Encapsulation," IEEE Electronics and Technology Conference, Orlando Florida May 29, 2001.

8. K. Takahashi et al, Tsukuba Research center, "Development of Advanced 3-D Chip Stacking Technology with Ultra Fine Interconnection" IEEE Electronics and Packaging Conference , Orlando Florida May 29, 2001.

9. E.F. "Jiggs" Giguere, Bell Laboratories, Holmdel, U.S.A., "Modem Constructions using folded flex circuit stacked assemblies," Internal report, Bell Laboratories, 1975.

10. W. Engelmaier, Bell Laboratories, "Designing Flex Circuits for Improved Flex Life", Proc. 12th Eletrical/Electronic Insulation Conference, boston, NA November 1975.

11. R. Aschenbrenner, Fraunhofer IZM, IEEE European System Packaging Workshop, Barcelona Spain January 2002.

12. H. Beutel, J-Uwe Meyer, et al, Fraunhofer IBMT, "Methods and Trends in Micropackaging of Active Biomedical Implants, IEEE Workshop on Chip-Package Co-Design, Zurich, Switzerland, March 2000

13. Thomas Harder, W. Reinert, Fraunhofer ISIT, Germany, "Low Profile Flip Chip Assembly Using ultra-Thin ICs", IMAPS 13th European Microelectronics ad Packaging Conference, Strasbourg, France June 1st 2001 p 310.

14. Vern Solberg, Tessera, "Stacked Multichip Packaging for the Next Generation Electronics, SMTA Pan Pacific Conference, Maui, Hawaii, February 2002.

15. J.F. Zeberle et al, Valtronic, Switzerland, "Flip chip with Stud Bump and Non-Conductive paste for CSP-3-D, Valtronic presentation IMAPS 13th European Microelectronics ad Packaging Conference, Strasbourg, France June 1st 2001 p 314.

Chapter 2

MULTI-CHIP CARRIERS IN A SYSTEM-ON-A CHIP-WORLD

Evan Davidson
IBM

2.1 INTRODUCTION

One of the key applications for foldable flex is to enable the use of 3-D (3-Dimensional) MCMs (Multi-Chip Modules). This embodiment of a 3-D electronic package (E-package) represents the ultimate in packing density and it should be used whenever it benefits the application. The use of any MCM, however, brings up the issue of SoC (System-on-a-Chip) versus SiP (System-in-a-Package). By its very nature, an MCM implies that the entire system is not on a single chip. Is eliminating all instances of an MCM what the SoC advocates mean when they so cavalierly bandy about the SoC acronym or can many SoCs exist on an MCM? Is the latter what the E-packaging community calls an SiP or is an SiP a more numerous collection of sub-SoCs (i.e., chips that are not SoCs) and/or SoCs on many single chip modules (SCMs) or one or more MCMs? These questions are very confusing and these questions are what will be explored in this chapter.

In its ultimate implication, SoC is the entire system on a chip. Conceptually, this means that eventually the semiconductor industry will attain the point at which an entire electronic system (no matter how large) can be placed upon a single chip. This may mean the use of many billions of transistors per chip. If one were willing to extrapolate Moore's Law as done by the 2001 International Technical Roadmap for Semiconductors (ITRS) out to 2016, one would find such a prediction.[1] If this prediction ever comes to fruition, what would be the role of the E-package in a future complex system? The answer is a minor one because all E-packaging would converge to a single chip on some kind of a substrate such as a single chip module (SCM) or chip-on-board (COB) with either a connector or direct surface mount, respectively.

Fortunately for the E-packaging industry, this single part embodiment for SoC is not likely to happen for the majority of applications because of the following factors:

1. Very large chip sizes are prohibitively expensive for large volume products because of yield and wafer productivity limitations. (Chip price typically scales with area raised to the 1.5 - 2 power.)[1]

2. Mixed technologies (CMOS, DRAM, SiGe, flash memory, analog and passive devices, etc.) on a single chip are also prohibitively expensive for large volume products because of process complexity and yield limitations.

3. Scalable systems require granularity; i.e., single chip designs with too much replicated content (e.g., many processor cores on a chip) may be inconsistent with entry system marketing goals because they are too expensive.

4. There is an inverse relationship between FET speed (device scaling) and on-chip metal scaling; i.e., very dense FETs require dense interconnections with smaller on-chip metal cross-sectional areas. The result is interconnections with poor electrical performance (high resistance (R), high capacitance (C), and high inductance (L) and/or wider-thicker lines with more levels of on-chip metal to compensate for the too small conductors.[1] This results in poor performance and/or high cost for complex metal layers on large chips with long lines.

5. The high cost of investment for modern semiconductor fabricators and the lower profit margins occurring in the very competitive chip business is causing many companies to rethink the advisability of building complex SoC parts with low yields because of the potential for insufficient return-on-investment (ROI). The high cost and competition factors are causing all companies to reconsider their willingness to continuously invest in the historically fast rate of advancing semiconductor technology.

All of the aforementioned items are deterrents for single chip SoC products. Item 1 has in effect been acknowledged by the 2001 ITRS. Previously, the 1999 Roadmap had been predicting continuously growing chip areas up to 900+ mm^2 (Year 2012) for low volume high-performance (as opposed to high volume cost-performance) microprocessors and ASICs (Application Specific Integrated Circuits).[2] In the 2001 ITRS, the corresponding value is a non-time varying flat 310 mm² from 2001 through 2016. This significant reduction in the predicted chip area is a result of some unexpected low yields being experienced with trying to fabricate 400+ mm² microprocessors in 2000 and 2001. It also recognizes that on-chip **functionality** is *suppose to double every two to three years* until at least 2016 which implies that there is new thinking that medium sized chips (12 to 17 mm on a side) will contain sufficient capability for realizing less aggressive but more profitable SoC designs essentially forever.[3]

Item 2 is another area where semiconductor manufacturers are finding that more complexity leads to lower yields and higher cost. For digital applications, system designers are specifying high speed digital CMOS logic and SRAM plus slower and denser DRAM all on the same chip. These chips tend to be in the 15 to 17 mm square size category because of the real estate required by the memories. In addition they require a few extra masking steps to add the DRAM process to the normal CMOS process for logic . The net result is a plethora of processing steps and difficult to produce parts that are too expensive for many applications. In the case of RF applications such as an SoC cell phone, there is a desire to place digital CMOS, analog CMOS, SiGe RF circuits, flash memory and passive devices all onto the same chip. There are semiconductor companies that believe they can do all of this but these embodiments for product are slow to happen because the resultant required investment and cost would be too high for the high volume low margin consumer business.

Item 3 brings up the point that many systems have variable features and/or are scalable in their multiplicity of function and that companies don't want to put a very expensive SoC into a product with all these "bells and whistles" if they can't sell it as a fully loaded product. From a marketing perspective, it makes sense to use many lesser function chips and to add more of them when the customer is willing to purchase a more feature-laden system.

Inherent in Item 4 is an argument that a very high device density on a chip drives a very complex multilayer metal interconnection structure which in turn negatively affects chip yield and cost. One of the inherent inconsistencies in Moore's Law is that the semiconductor device structures and metal structures track in opposite directions; i.e., as devices get denser with better performance, on-chip metal interconnections must also get denser but they exhibit poorer performance. The problem is that smaller, closely packed lines demonstrate higher resistance, capacitance, and inductance. High frequency phenomena such as skin effect, crosstalk and lossy dielectrics make these problems even worse. So we have the situation where device people want to run 30 GHz local clock rates (Y2016) with ten levels of metal chip interconnections to allow for the use of many levels of fat wire (wide and thick metal) for long lines.[1] Assuming these complex large chip structures are feasible, they will be very expensive and their applications will be limited in number.

With 20 nm lithography's predicted ability to produce billions of transistors per mm^2 in 2016 and densely packed small cross-sectional lines with hundreds of picoseconds (ps) per mm propagation delays, there is a glaring inconsistency in reconciling the FEOL's (Front-End-Of-the-Line: the transistor devices) capability with the BEOL's (Back-End-Of-the-Line or levels of metal) excessive propagation delay. Clearly there must be a solution to this problem and the answer is to use many low density fat wire metal layers with adequately fast conductors. At 30 GHz, this will require more than copper conductors and low dielectric constant (ε_r) insulators. To achieve no more than a few ps/mm propagation delay, virtually lossless transmission lines with minimal crosstalk will be required and this requires lines and separations with many micrometers (μm) dimensions. For a 20 nm semiconductor process (Y2016), this mismatch between device dimensions and metal dimensions is not a good situation. To effect a purely semiconductor solution, requires many low density obese wire layers (dimensions in microns when lithography is sub-100 nm) with many additional levels of metal (LOM) to provide adequate chip wiring capacity. As an indication, the 2001 ITRS predicts ten LOM beginning in 2010. (This is relative to 7 LOM in 2001.) Achieving the required new inter-metal dielectrics ($\varepsilon_r \leq 2$) and the additional LOMs is one of the listed Grand Challenges (or planned inventions) listed for beyond Y2008.

Item 5 brings up the business case status for semiconductor investment. It says that new fabs are expensive, competition is keen and margins are low. This does not constitute a good environment for taking high risk to aggressively develop new technologies. Perhaps there will be competitor fall-out and profit margins among the survivors will rise in the future. Even if this were to happen and the high costs of overly aggressive SoC designs were an acceptable risk, the high chip prices that would result are not good for a high volume consumer market. Historically, the semiconductor business has always required high volumes to be viable because of the large investments involved so if all the economics don't work out properly for SoC, it will at best be only partially successful (for low volume parts) and at worst a failure.

Packaged electronics at the systems level encompasses both chips and packages; i.e., E-packaging. In fact, E-packaging is the place where technology optimization should occur. As discussed above, the road to a single chip SoC is fraught with all kinds of pitfalls. Perhaps better technology and cost optimization will occur with multi-chip SoCs and/or non-SoCs constituting the product.[4] By accepting this assertion as a possibility, the reader should now accept that there is less than a trivial role for E-packaging than just SoCs on SCMs and COBs in the future. In fact, it appears that various forms of MCMs will be very important in the future. The single chip SoC may exist in some products but it is not the wave of the future for the majority of products.

In the 1999 ITRS, there was a special section devoted to SoC. This is not the case in the 2001 ITRS. The absence of an SoC section may indicate that the original vision for the once omnipotent SoC is being downplayed by the semiconductor community. They are, however, stuck with a packaging related dilemma caused by the inability to do everything with SoC that they don't like. It is articuiated by the following quote from the 2001 ITRS:[5]

"If future semiconductor products must be targeted to maintain constant

or decreasing prices and the average number of pins per unit increases at 10% while the average cost per pin decreases at only 5%, then the following will occur:

1. the average packaging share of total product cost will double over the 15-year road map period, and

2. the ultimate result will be greatly reduced gross profit margins and limited ability to invest in R&D and factory capacity.

This conclusion is one of the drivers behind the industry trends to

reduce overall system pin requirements by combining functionality into System-on-Chip (SoC) and through the use of multi-chip modules, bumped chip on board (COB), and other creative solutions."

This quote is interesting in many respects. Firstly, we see the well-known fact that packaging cost is proportional to the total number of signal and power pins. Next is the implication that total pin count will increase with time because of the ever increasing on-chip function and power requirements but pin cost will not reduce at the same rate as function cost. This is shown in Figure 1. Consequently, the total cost of packaging as illustrated in Figure 2 becomes too expensive and it hurts the entire industry. Now the most interesting part: *SoCs are no longer the total solution and MCMs and COB may be part of solving the future cost dilemma.*

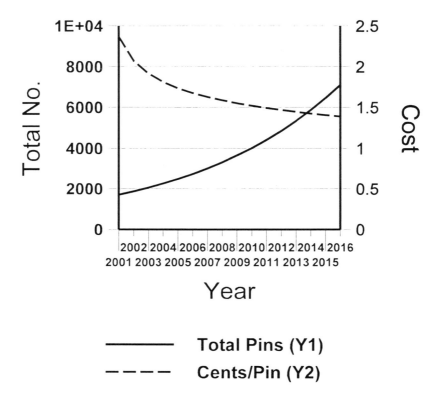

Figure 1. Trends for total package I/O and per pin cost.

What the author of the quote appears to be saying is that, in at least some applications, the semiconductor industry may have to optimize cost at the packaged electronics level rather than only at the chip level. This is a healthy conclusion that basically says that chips and at least first level packages should be treated as an integrated product from the business case perspective. In also says that custom chip and ASIC design companies have to be tightly integrated with packaging manufacturers. Merely putting an expensive high margin chip on a low-cost SCM with a few hundred solder bumps or pins will not be sufficient in the future. Although the 2001 ITRS Executive Summary doesn't explicitly mention it, the Assembly and Packaging Chapter does begin with the statement: "There is an increased awareness in the industry that assembly and packaging is an essential and integral part of the semiconductor product." This indicates that the semiconductor people who, heretofore, dominated the determination of packaged electronics systems with packages playing a lowly role are finally realizing that packages are not merely mechanical carriers; rather they are an integral part of the design of a high performance system.

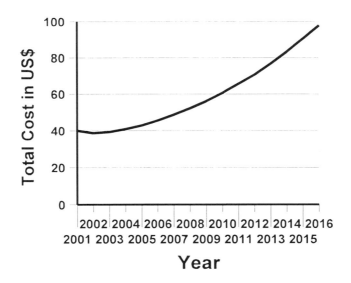

Figure 2. Trend for total cost of a high-end package. (Source: 2001 ITRS)

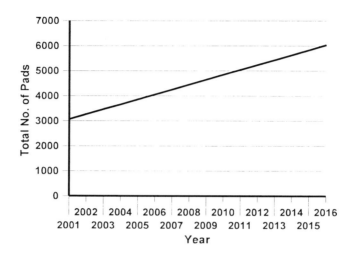

Figure 3. Trend for total I/O on a high-end chip. (Source: 2001 ITRS)

2.2 FUTURE PACKAGES

The 2001 ITRS defines packaging totally in terms of SCM pins required per chip. According to the ITRS, the stress case for packaging is the high-performance ASIC. A plot of their projections for ASIC pins/chip and pins per SCM are given in Figures 3 and 1, respectively.[6] Other very important chip requirement parameters that affect E-packages are chip performance (MHz.) and chip power density (W/cm^2). These are shown in Figure 4.

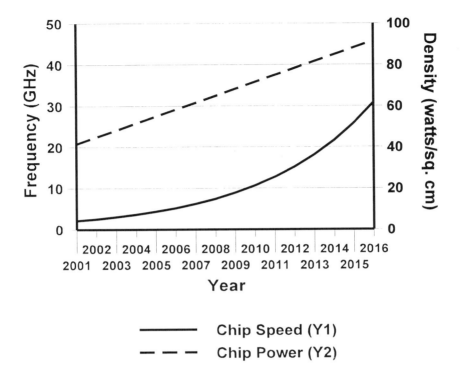

Figure 4. Predicted clock frequency and power density for a high-end chip. (Source: 2001 ITRS)

From Figures 1 and 3, we see that high-end ASIC (and microprocessor) chips and packages will have thousands of total I/O. The high numbers shown are caused by signal pins increasing due to high bandwidth buses with many lines and multi-port switching elements in computers and networking equipment. Total I/O also increases because very high frequencies require low package power loop inductances, low signal coupling inductances and low power distribution resistances. Low inductances and low resistances cause the number of power and ground I/O to increase to provide good signal shielding and good power distribution. (In fact a signal-to-power ratio of one is assumed in the 2001 ITRS for high performance ASIC packages.) All of this means that E-packaging must continuously improve line and via density, support many power layers and be thin. For good ultra-high frequency signal propagation, dielectric materials with low permitivitty ($\varepsilon_r \leq 4$) and low loss ($\tan\delta < 0.001$) are required. Furthermore, very low thermal resistances from the chips to cooling air or liquid are required because of the high powers associated with future products.

For high performance ASICs and microprocessor SOCs, MCMs come into play as a means to reduce total system cost because they can reduce the total number of connections to the second level package (cards and boards). This happens because MCMs coalesce the preponderance of total chip I/O into the first level substrate with intra-module wiring. This significantly reduces the number of first-to-second level interconnections (an important part of the price for packaging); thereby, reducing total packaging cost. Once the resulting lower complexity and cost of the second level PWB (printed Wiring Board) package is considered, the additional cost of an MCM (vs. an SCM) may be offset in many applications.

The other set of requirements for packaging occurs in products where miniaturization is required. Examples are cell phones, implantable devices like hearing aids, intrabody diagnostic medical devices and personal data assistants (PDAs). These products are characterized by less I/O requirements, low power and high volumetric packing density. They are also the products that require mixed technologies. As an example, a cell phone requires RF circuitry such as GaAs or SiGe, analog CMOS, digital CMOS, DRAM, flash memory and passive devices. Each of these process technologies are different and it would be very expensive to combine them all onto one SoC. A much more sensible approach would be to use a stacked chip 3-D MCM fabricated with foldable flex or a dense planar MCM. For these packages, each chip would be small, probably thinned and made out of a different process technology. Resistor, capacitor and inductor passive devices could also be placed on the flex films.[2]

In the 2001 ITRS, the miniaturized packaged products are categorized as the "Hand-held" market. In this sector, projected chip area is less than 90 mm², package thickness is less than 0.5 mm and the power per chip is less than 3.5 watts. The total Hand-held package I/O varies with time as depicted in Figure 5. Note that even though hand-held products are portable, Figure 5 still predicts a significant amount of I/O will be required for the most complex chips used in these applications. High I/O and the use of mixed technologies again may make the use of MCMs a cost-effective solution for the Hand-held market.

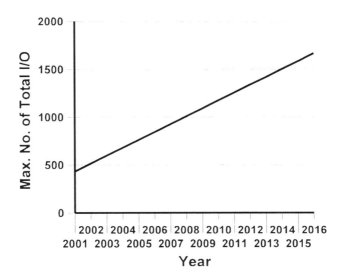

Figure 5. Trend for total package I/O for a Hand-held package. (Source: 2001 ITRS)

There are two basic material sets that have been used for fabricating E-packages: ceramics and organics. Historically, multi-level ceramics (MLC) have been used for complex high I/O first level packages including SCMs and MCMs. Low temperature co-fired ceramic (LTCC) substrates are also available with integrated passive elements for telecommunications applications. In the past, organics had been used for simple (e.g., low I/O flat packs) first level plastic modules and printed circuit cards and boards. However, in recent years organic structures have begun to encroach upon the applications space enjoyed by MLC. This occurred when high density built-up thin film wiring layers were applied to PWB (Printed Wire Board) cores enabling high I/O (greater than 250) single chip modules for microprocessors.[3] Introduced by Intel in their Pentium II line of mobile microprocessors in the late 1990s, organic modules are now becoming prevalent in many application sectors for E-packaging. Flexible films, the subject of this book, are considered an organic technology that is likely to become more pervasive in the future in its tape and laminated forms.[4]

The on-going debate about SoC vs. SiP in the early '00s is caused by the cost associated with trying to fabricate large and complex microprocessor and

ASIC chips. If it were feasible to use multiple smaller chips on an SiP rather than a single chip at lower cost without any loss of function or performance, then the SiP approach would make sense.[5] The latter implies that MCMs could become more prevalent in the future than one might expect.

MCMs have been around since the early 1980s. They were introduced by IBM in a mainframe computer as a means to reduce multi-chip path delays.[6] Another benefit for the use of MCMs when compared to SCMs on a board is the reduction in board complexity by bringing the inter-SCM wires onto the MCM. By the late '80s, other mainframe and supercomputer companies began to use MCMs. They didn't, however, catch on for mainstream applications such as workstations and PC (Personal Computer) servers until the mid-90's when Intel introduced the three chip Pentium Pro microprocessor MCM with two satellite (back door) L2 (Level 2) SRAM (Static Random Access Memory) chips.[7] This module was fabricated on a ceramic substrate with thin film wiring (TFW) used to connect the microprocessor to the cache chips.

Until the early '00s, MCMs were mostly made using MLC technology. As was the case for the Pentium Pro module, some of these MLC substrates included layers of organic TFW for increased wiring capacity and EC (Engineering Change) and repair capabilities.[8] MLC MCMs, however, were never seriously considered for high volume computers because they were too expensive for the consumer market. MLC processing is fundamentally more expensive for high volume modules than organic processing because ceramic green sheets sizes are smaller than organic panels. This causes fewer batch fabricated MLC parts per production run relative to organic parts. In addition MLC involves the screening of high temperature refractory metal pastes for conductor paths and mechanical punching for layer-to-layer vias. This is in contrast to the photolithographic etching processes used for forming wires and the mechanical drilling, laser drilling or etching used to make organic vias. As a result of these fundamental differences in processing, MLC parts can not be made in the same factory as organic parts. Since the organic PWB business is much larger than the MLC business, very few organic packaging companies were ever willing to invest in the ceramic business. With high investment requirements and limited competition in the MLC business, prices for simple ceramic SCMs were never low enough to displace the early plastic (organic) packages.

By the beginning of the new Millennium, organic packages began to exhibit higher wiring and I/O density characteristics. Current constructions can be as high as five built-up TFW layers on the top of the substrate and five on the bottom sandwiching a four power layer core with mechanically drilled plated through hole (PTH) vias (a 5+4+5 structure). The latest embodiments of this type of module are using laser drilled (or etched) vias called microvias for even higher interconnection densities. These new organic modules can be either SCMs or MCMs and they are amenable to high volume batch fabrication techniques. It is now possible to build a complex organic module with embedded passives out of laminated flex layer technology using these modern fabrication techniques. For low power chips, the flex can be folded to form 3-D modules with thinned stacked chips.

2.3   FUTURE SYSTEMS TRENDS

The 2001 ITRS lists the following product categories and definitions for E-packaging applications:
- Low-cost -- less than US$300 for consumer products, micro-controllers, disk drives, displays.
- Hand-held -- less than US$1000 for battery powered products such as: personal data assistants (PDAs), cellular phones and other hand-held products.
- Cost-performance -- less than US$3000 for notebooks, desktop computers and low-end network equipment.
- High-performance -- greater than US$3000 for workstations, servers, avionics, supercomputers and high-end network equipment.
- Harsh -- Under the hood and other hostile environments.
- Memory -- DRAMs and SRAMs.

In the early 2000s, Low-cost packages were mostly surface mount plastic flat packs. In the future, inexpensive organic chip scale packages (CSP) using microvia laminate substrates and microvia boards are likely to be prevalent in this category for small chips with high I/O counts. For future Hand-held and Cost-performance products, there are many organic candidates such as thin dense CSPs, MCMs, 3-D folded flex modules and laminated flex structures with passive devices. The Harsh sector with high temperatures and corrosive atmospheres is where ceramic substrates will continue to be used. Finally, plastic flat packs and 3-D folded flex packages will be the choices for the Memory category.

For E-packaging strategic purposes, the High-performance and Hand-held application areas tend to drive the most important advanced development activities. High frequency operation, high power, high I/O counts, and high wiring density are the attributes of the High-performance sector. These items drive advanced materials and processes and advanced cooling requirements. On the other hand, the Hand-held sector drives low power designs, embedded passive materials and processing, plus miniaturization. The new developments for these two categories tend to cover the other categories except for the Harsh category. The latter is where MLC (which can be hermetically sealed) is likely to continue to be used. Since MLC technology is relatively mature for complex applications already, not a lot of new advances are expected in this area.

2.4   FUTURE FOR THE HIGH FREQUENCY SECTOR

No contemporary semiconductor company is more dependent upon high-frequency packaging than Intel. Since the advent of the microprocessor in the early 1970s, Intel's marketing strategy has been to sell clock frequency. Until the mid-nineties, their microprocessor frequencies were below 300 MHz and they were able to purchase commodity ceramic PGA (Pin Grid Array) SCMs from independent suppliers. However, beginning in the early '90s, Intel began to realize that they had to be able to influence the course of advanced packaging to control their application specifications and cost to manage the success of their main microprocessor business. As a result, they created an internal packaging program to develop designs and processes to meet their needs. Rather than investing in their own packaging manufacturing facilities, they worked with various Japanese companies as manufacturers of Intel designed components.

Once Intel entered the E-packaging design business, they chose the built-up organic package as a cost-effective high frequency package.[7] They called it an OLGA (Organic Land Grid Array) package and it came with a PGA or BGA (Ball Grid Array) module-to-board connector.[4] A picture of the PGA version is shown in Figure 6. As you can see, the OLGA module has a PWB core with thin film built-up layers on the top and bottom surfaces of the core. A pinned interposer is soldered to the LGA pad pattern on the bottom of the substrate for a pluggable connection to the motherboard. Flip-chip connections are made to the top of the substrate.

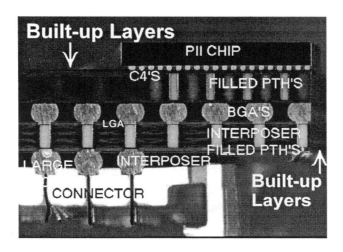

Figure 6. Photograph of the cross-section of an Intel organic land grid array (OLGA) SCM package.

In 2002, a variant of OLGA called FCPGA (Flip Chip Pin Grid Array) became the main packaging technology used by Intel. (The main difference between FCPGA and OLGA is the pin-grid-array is now directly soldered to the substrate without the use of a land-grid-array BGA connection to an interposer with staked pins as shown in Figure 6.) This type of built-up substrate has become pervasive in many other applications used by other companies and it is now the preferred low-cost, high-frequency, high I/O package.[8] It has essentially displaced MLC SCMs in most of this market. Until recently, FCPGA was either an SCM or FCM (Few Chip Module) with a microprocessor chip and/or L2 cache chip(s). A good example of the latter is the Intel Itanium 1 cartridge shown in Figure 7. (Note that this is an OLGA microprocessor SCM with a four chip OLGA L2 FCM on an OLGA card.) The cartridge has a PGA connector on the bottom for signal connections (with power pins for return paths and crosstalk reduction) and an edge connector for the main power current.

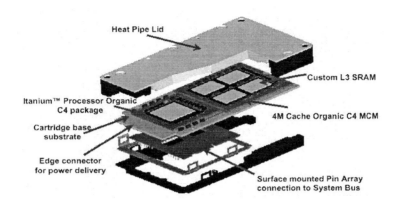

Figure 7. The Intel Itanium 1 cartridge package. (Source: Intel Corp.)

At times Intel recombines the separate L2 chips with the microprocessor chip when the next denser semiconductor technology becomes available. This is done to reduce costs by gravitating towards SoC whenever it becomes feasible. For example, Intel's next generation Itanium 2 product (code named McKinley) is reputed to be an SCM with the five chips in Figure 7 coalescing into one to reduce the packaging and overall product cost. This is accomplished by switching from a $0.18\,\mu m$ to $0.13\,\mu m$ generation semiconductor technology.[9] This scenario of bringing out a new product with multi-chips and reducing them to a single chip when a better production technology is available is commonly thought to be a way to improve performance, reduce power dissipation and diminish cost. This scenario has been a key driver towards making SoCs. The potential flaw in this approach is that new higher density chips have more complex process technologies and they may have to be sufficiently large to contain the expanded function resulting in a high total defect susceptible area such that low yields and relatively high costs for SoCs are likely to occur.

As semiconductor ground rules inexorably improve from ITRS node to node and operating frequencies increase beyond 2 GHz., another concern for large chips is the smaller cross-sectional areas of global lines.[10] There is an irony here that has been mentioned before: as devices and circuits get faster, long global line delay increases because of higher R, L, and C. Call this a reverse scaling effect. The solution to this problem is lower ε_r for BEOL insulators, thicker dielectrics, wider lines and larger line-to-line separations. The result is even more layers of metal and greater chip cost. This is the underlying incentive for substituting smaller chips on a SiP for a large SoC; i.e., multiple chips that are less complex are cheaper to make than a sub-optimized overly large chip.[9]

This concept of building an FCM microprocessor system that is subsequently reduced to an SCM when denser semiconductor technology becomes available has worked well up to the 0.25 μm semiconductor node. As CMOS technologies progress beyond this point with levels of metal at seven and increasing up to 10+ while the device and on-chip wire critical dimensions becoming smaller, chips keep getting more difficult to manufacture. As total on-chip defect susceptible areas increase, yields for very large chips (>300 mm^2) became smaller. Instead of the customary 50+% yields for high volume chips, they began to hover around 10 - 20%. As a result of this and greater competitive pressures, gross profit margins will decrease significantly for these very large complex ASICs and microprocessors. This is already happening and the business men are busily looking for ways to reduce cost.

If overly large chips with excessive defect susceptible area are part of the problem, then perhaps smaller chips on a FCM with equal function and performance at the packaged electronics level at a lower overall cost is a better way to build a system. Many people have reached this conclusion based upon business case studies and this is why the concept of SiP is becoming more prevalent.[5] As explained above, the smaller chip sizes and the indication that MCMs have a role in the future as mentioned in the 2001 ITRS, appears to bear this out.

2.5 IDEAL HIGH PERFORMANCE PACKAGE

Many types of off-chip applications can stress the requirements for a high-performance package. One of them, which is perhaps the greatest challenge, is a high-end processing cluster for a large SMP (Symmetrical Multi-Processing) server. An illustration of one of these clusters is shown in Figure 8. Many different companies including Intel are and will be interested in building this type of parallel commercial server in the near future. Uses for this type of machine are: web server consolidation, numerically intensive scientific calculations, large database management, large batch processing loads, multiple operating systems operating as logical partitions, etc.

Figure 8 depicts a central electronics complex (CEC) consisting of a total of 16 processors on four chips with local and global system control and cache memory with sufficient interconnection bandwidths to keep each individual processor running at a very high utilization rate. Non-blocking crossbar switch fabrics are used such that communication paths are always available between each processor, main memory and external I/O. Connections to the I/O sub-system use high speed serial source-synchronous phase aligned unidirectional busses. Each of the 2 differential DDR (Double Data Rate) I/O links simultaneously transfer one byte of data in two directions at 5-10 Gb/s. rates. (Typical wire count per bidirectional link is 40 wires including clocks and control lines.) The CEC cache coherence function is performed by the L2.5 directory located on the central system control trip.[11] Finally a 4-way high bandwidth interleaved main memory (DRAM) is used to provide large amounts of data to the CEC.

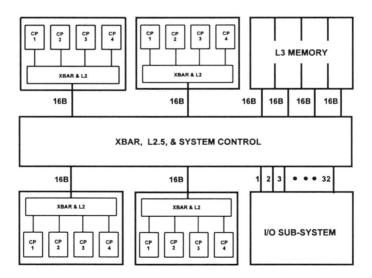

Figure 8. A high-end digital system designed to illustrate the trade-offs between SoC and SoP.

The CEC, L3 main memory and I/O sub-system shown in Figure 8 clearly falls into the ITRS High-performance Sector. As defined, there is a large amount of I/O per chip and a great amount of interconnections between the chips. Each 4-way central processor (CP) chip could be considered an SoC. One could also say that the central system control (SC) chip is also an SoC. Hence, It is possible to treat a complex cluster system as a collection of SoCs. This is probably a more reasonable way to interpret the meaning of an SoC: rather than assuming it means an entire system on a single chip. A logical way to package the contents of Figure 8 on an MCM will be described subsequently.

One point to note about the four processor CP chips contained in Figure 8 is that these chips can be useful if only one processor per chip and the crossbar plus part of the L2 (with or without redundancy) on the System Control chip are operational. This is a way to use partially good chips to improve effective yield and provide fewer processors for a lower-end product by using what can be defined as a single SoC design. Since the CP chip is made up of five small functional cores, even the long lossy line concern may be acceptable. SoCs designed with parallelism and/or redundancy that can be used as partially good die can be economically viable even if they are very large.

The attributes of a leading edge High-performance package are:
1. High density multilayer wiring capability.
2. High density I/O capability.
3. Low resistance conductors.
4. High power distribution current capability.
5. Thin dielectric layer thicknesses for superior AC power and signal distribution.
6. Ability to achieve very low loop inductances for decoupling capacitors.
7. Ability to cool very high power density chips.
8. Ability to support multi-chips for SiP applications.

Item 1 implies that signal and power conductors need to be defined by photolithographic or direct write raster technologies. Optical masks or scanning e-beams or laser beams with exposure sensitive etch resist or ablation technologies are examples of such high density wire making technologies. High density I/O capability as mentioned in Item 2 use the same kinds of processes to form vias and pads for joining chips and areal module connector elements such as studs, solder joints, pins and land-grid-array (LGA) elements. Items 1 and 2 preclude the screening and punch technologies commonly used in MLC structures because they aren't sufficiently dense.

Item 3 implies that pure metals rather that more resistive metallic pastes should be used. The latter are common in MLC structures. For both first (modules) and second (PWBs) level organic packages; copper, which is essentially the best conductor, is generally used.[12] For Item 4, Items 1-3 are prerequisites because the requirement for low voltage drop and high currents mandates a multitude of parallel low resistance wires, vias, and I/Os.

Thin dielectric layers (Item 5) in a high performance package are required to achieve a well-controlled characteristic impedance for a signal transmission line conductor that is narrow (for high density). For power distribution, a small separation between supply and return currents (lines and planes) is required for the low loop inductance needed to support high frequencies and high switching currents. The total thickness of the package substrate is also important because for the best power distribution system, currents should flow predominantly vertically through closely spaced supply and return vias with a short current loop length (lowest effective inductance [L_{eff}] for the package). L_{eff} is also affected by the number of parallel current paths in the power distribution system. This, too, requires additional vias. A small L_{eff} results in low switching noise on the chip power supplies to assure proper function for circuit operation. Besides thinness, low ε_r for each interlayer insulator is also desirable because it improves signal propagation delay and reduces crosstalk.

Item 6 is an important additional means for reducing package L_{eff} and switching noise. To accomplish this, high frequency (low inductance) chip-type ceramic capacitors are placed on the module substrate as close to the chip as possible. If placing them on the bottom side of a thin packaging substrate directly underneath the chip is possible, this would be the best location. These capacitors do not filter the noise at the actual chip switching frequencies, only on-chip capacitances can do that, but they do provide a mid-frequency mechanism for recharging the on-chip capacitances. The faster this recharging can occur, the smaller the average band of superimposed high frequency noise on the chip's power and ground voltage rails.

From Figure 4, one can see that chip powers will be over one hundred watts for some high performance parts.[13] The power distribution requirements for such high powers have already been discussed but with these elevated powers comes cooling requirements. There are many techniques available for cooling high power chips; some involve high velocity air cooling and some involve liquid cooling. All high power cooling techniques require direct access to the back side of the chip. This essentially means that for Item 7 to be satisfied, only flip chip package bonding techniques are permissible for the most demanding applications. Another factor to consider is that partitioning an SoC function into multiple chips that are spread out on an MCM reduces the average power density. Since cooling systems are stressed by power density more so than total power, the SiP can be used to reduce the expense of the required cooling solution.

Item 8 deals with the anticipation that SoC will not be an optimal solution for most complex system applications as explained above. However, to achieve the performance goals for typical applications, SiPs utilizing MCMs will be required. Consequently, any desirable future high performance packaging should support multi-chip capability to keep interchip wire lengths short and dense with transmission line discontinuities kept to a minimum. The latter attributes are required for high frequency, high bandwidth and low latency system design. Relative to using SCMs for an SiP, MCMs are often more desirable because they reduce the total I/O that a PWB has to support by absorbing the chip-to-chip I/Os on the MCM such that they do not reach the first level package PWB connector. Since E-packaging complexity is driven by total pins supported, the PWB's cost is reduced. This is especially important in the case where a few high density-high I/O SCMs that occupy a relatively small area of a large board would have caused the entire PWB to be very complex and expensive. It is for this reason that the total chip and E-package should be optimized at the system level to assure that all the proper cost tradeoffs are being made. Just getting the lowest competitive price for each separate component may not lead to overall system optimization.

Items 1-8 describe the requirements for a leading edge high performance package of the future. A company that requires this capability for their microprocessors and chipsets is Intel. Recently, Intel released details about a future packaging technology which they call BBUL (Bumpless Build-Up Layer) packaging.[10] By studying the attributes of Intel's announced development program for BBUL, one can see that they have addressed all of the issues mandated by the above items.

2.6 ANALYZING THE BBUL PACKAGE

BBUL integrated chip packages are depicted in Figures 9 and 10. Its basic feature is that through the use of built-up thin film application techniques, the SCM or MCM packages are totally integrated with the chip(s) or, alternately, the chips are impregnated into the package. In simplistic terms, cutouts are made in a solid organic core material (BT – Bismaleimide Triazine) that is equal to the thickness of the chips. After processing up to their pad level, the chips are placed into the cutouts and glued in with an adhesive material. Once the chips are in their insets, a thin film wiring layer is stud bonded to the upper pad layer of the die. Subsequently, more built-up layers are added for inter-chip, intra-chip, fan-out and module I/O wiring. Module I/O can be either pads (LGA), pins (PGA) or solder balls (BGA). In the BBUL package, the chip backside is exposed for enhanced conductive cooling purposes. Also note in Figure 11, decoupling capacitors can be placed directly beneath the chip power terminals for superior switching noise reduction.

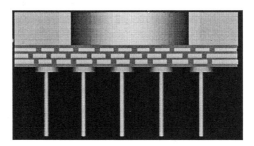

Figure 9. A cross-section of the Intel BBUL package to illustrate the concept. (Source: Intel Corp.)

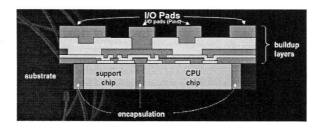

Figure 10. Details for a multi-chip BBUL package. (Source: Intel Corp.)

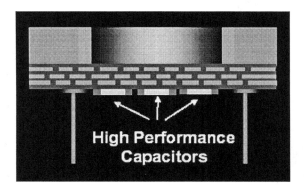

Figure 11. Placement of decoupling capacitors on the BBUL package. (Source: Intel Corp.)

Now why feature a purely built-up package in a book about a flex packaging? There are multiple reasons for doing this. One is to illustrate a concept of an excellent approach for the design of a future high- performance, high-density, high-power package with SiP characteristics as indicated in Figure 12. It also appears to be amenable to large panel processing with many parallel parts. This implies that the part could be relatively inexpensive. The aforementioned Items 1-8 are well represented in the definition of BBUL. Many experts, however, believe that the challenges that Intel will incur to make this novel approach into a high volume manufacturable integrated chip-package are quite formidable. (This is not meant to imply that Intel, with its great resources and commitment, will not make BBUL into a commercial success.)

In the case of BBUL, all the chips have to be known good die (KGD) before they are placed into the core cutouts. Also, there are placement precision and CTE (Coefficient of Thermal Expansion) materials mismatch concerns. Finally, the yields associated with each serial step of this chip-in-package (CiP) construction must be very high. This is especially true because both interim and post fabrication repairs may be impractical or unaffordable. In the case of a multi-chip BBUL, an imperfectly processed part might have to be discarded making the effective cost of the good ones undesirably expensive. Nevertheless,

38

it is likely that Intel has done their technical homework and with the need to make BBUL viable for future product, they are motivated to use their formidable resources to make this electronic packaging concept an economically viable approach.

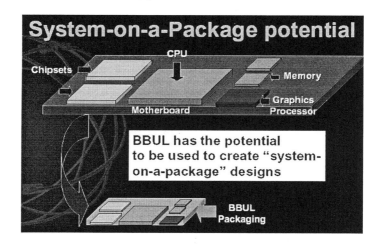

Figure 12. The concept of an SiP using the BBUL package. (Source: Intel Corp.)

If we are receptive to accepting the BBUL concept as a desirable product but assume that it is unaffordable by companies other than Intel, there may be processing techniques that are already more developed and more mainstream that can compete effectively with Intel's approach. This is where flex technologies can be considered.

One of BBUL's less than desirable attributes is the sequential and difficult to repair nature of the built-up TFW process. Consider starting with high cost KGDs and adding the BBUL layers to form the packaged die(s). Any defect in the BBUL process could force the whole valuable assembly to be discarded (assuming the structure is non-repairable). So if each per step processing yield for BBUL is below 95%, final test yields of 75% (for a six step process) are the result. Discarding one in four finished SCM or MCM assemblies, would be a costly penalty. So for Intel to make BBUL commercially viable for high volume microprocessor parts, they will have to achieve very high processing yields and/or low cost repairability.

2.7 ALTERNATIVES TO BBUL

Fortunately, there are other ways to achieve BBUL-like structures using flex techniques where the thin film layers are built-up on a glass substrate separately and tested before being joined to the chips. Now KGD can be mated to KGTFW (Known Good Thin Film Wires) and only the joining step has to have high yield. IBM has developed such a process and they call it TFT-J (Thin Film Transfer-Join).[11]

Flex TFT-J was originally developed for MLC substrates as a way to increase wiring capacity by adding topside multi-layer TFW layers.[14] TFT-J was also introduced to replace the higher cost built-up TFW layers IBM had been using on the large MLC MCMs used in their mainframes. In addition, TFT-J has been applied directly to precision aligned chip arrays or macros (called PAM) with stud connections as a way to construct an SoC CiP (Chip-in-Package, a variety of SiP) that is very similar to BBUL in function, performance and construction as shown in Figure 13.[12]

In another instantiation of TFT-J, the multi-flex layers can be applied to a laminated microvia flex substrate as also shown in Figure 13. This exhibits the more standard process of joining separate chips to separate packages using a high density flip-chip solder bump technology. Chips placed on a microvia flex laminate substrate with either TFT-J or built-up layers can rival BBUL with respect to performance and function. TFT-J can also be cut into small patches and placed only onto the regions of the substrate where high-density I/Os occur to reduce cost. In this usage, the TFT-J built-up flex layers can be used as a localized fan-out space transformer. The TFT-J construction technique is shown in Figure 14.

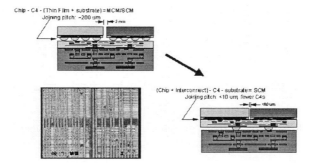

Figure 13. The structures for TFT-J built-up layers showing its use on a package substrate or it use as directly applied to chips for joining and wiring discretely partitioned SoC functional chiplets. The photograph depicts actual hardware with wiring crossing the inter-chip gaps. (Source: IBM Corp.)

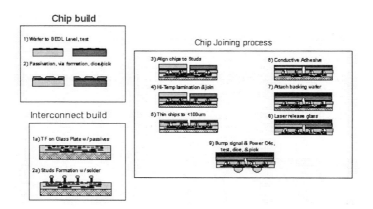

Figure 14. TFT-J processing: the chips (or packaging substrates) and thin films are processed in parallel and subsequently joined. The thin films are built-up on a sacrificial glass carrier and joined to chip studs. After chip thinning, a silicon back support is added and the glass carrier is released with a laser process.

For less demanding wiring density and I/O count applications, the TFW can be left off of the microvia core. This results in even lower cost and, with or without TFW, the entire MCM can be tested before KGD joining occurs. This will improve final test yield and manufacturing costs. If justified, roll-to-roll high volume batch flex fabrication for the microvia core could be added to the manufacturing process and the result is a common way to build organic modules, cards and boards using the same technology in a single factory.

2.8   HIGH-END PACKAGING EXAMPLE

At this point, it would be instructive to look at how one might want to package the multi-processing system shown in Figure 8. This system consists of the following inventory:

-    Four CP chips with 4 processor cores per chip and a crossbar switch fabric.
-    One system control switch chip with centralized coherent L2.5 cache, a main memory controller and an I/O controller.
-    Four L3 3-D flex DRAM stacks.

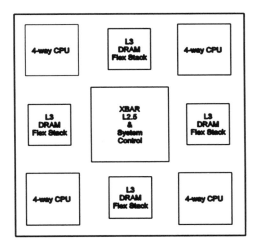

Figure 15. Chip layout of an MCM for the computer server system shown in Figure 8.

A plan view of an MCM containing this function is shown in Figure 15 and a cross-section of this module is shown in Figure 16. The core of this module is made from a laminated flex technology with micovias either laser drilled or etched. The pitch of these vias is the same as the C4 (Controlled Collapse Chip Connect) solder bump connections on each chip and the memory 3-D stack. The module to board connector can be any type; e.g., surface mount BGA (Ball Grid Array), CGA (Column Grid Array), separable LGA (Land Grid Array), or pluggable PGA (Pin Grid Array). Decoupling capacitors are shown placed on the bottom of the substrate in areas where the module-to-board connections are absent. The card/board technology used can be the same as the module substrate technology; i.e., laminated microvia flex layers.

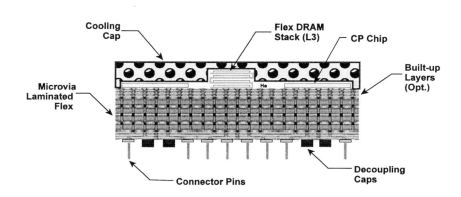

Figure 16. Cross-sectional structure for a microvia laminated flex MCM for the chip set shown in Figure 15.

The just described packaging technology has all the attributes of an advanced high performance package as described above. It can handle high I/O, high wiring demand, high speed, high power and high cooling capacity. It is basically a pervasive multi-layer flex technology with 3-D folding capabilities that can be both the first (module) and second (PWB) level packaging. If roll-to-roll processing is used, there is a single factory economy of scale involved that could make microvia flex laminate a high volume, low cost and high function technology.

2.9 TELECOMMUNICATIONS PACKAGING

The applications for the laminated flex technology described so far are predominantly digital. For telecommunications; RF, analog, digital and passive technologies are all required. A packaging technology that can support the simultaneous existence of all these technologies is very desirable. Fortunately, with some enhancements from what has been described, the microvia flex laminate package can do the job.

The University of Arkansas in conjunction with the Sheldahl Corporation has developed an ideal organic package for advanced telecommunications packaging. [2] Note from Figure 17 that special layers are included for high value resistors (CrSi - 200 Ω/$\square$) and high value capacitors (Ta$_2$O$_5$ - 200 nf/cm^2). Inductors can be either the single layer spiral type or the multi-layer helix type. Because these structures are buried in the substrate, their parasitics are very low and high Q factors (>80 @ 2 GHz) are achievable.

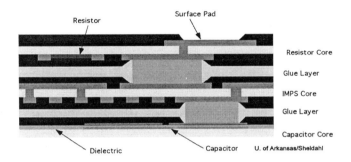

Figure 17. Cross-section of a microvia laminated flex structure with integrated resistive and capacitive layers.

The laminated substrate shown in Figure 14 was developed for a microvia roll-to-roll processing line. Since this type of fabrication is integrated and consistent with very high volume production, substrates made this way should be inexpensive. Placing small single technology chips (e.g., GaAs, SiGe, analog, digital, flash, and DRAM) on this type of MCM substrate with passive devices does have the potential to be an optimal low cost SiP solution for all kinds of mixed RF, analog, flash, and digital technologies.

2.10 SoC - SoP TRADEOFF

There are applications where single chip SoC will prevail and there are applications where multi-chip SiP will prevail. Taking all factors into account (performance, wafer size, chip yield, process complexity, fab investment and package cost), the ultimate determiner of a chosen design point should be a cost analysis and risk analysis of the feasible options when the function and performance for both the SoC and SiP embodiments meet the application's requirements.

Because of the on-chip long lossy line effect, higher performance will drive even more levels of metal than dictated by just the higher circuit density requirements; thus, making large semiconductors more and more complex and costly. Telecommunications applications will also become more costly because of the need for mixed technologies on chips and greater semiconductor complexity plus the need for integrated passive components. Finally, the microvia laminate organic packaging industry will become more mature with large panel and roll-to-roll processing causing packaging prices to fall. As a result, trading off a higher cost chip for lower cost chips placed upon low cost packages, such as organic flex FCMs and MCMs, will become quite attractive. With the wireless industry driving the need to integrate discrete devices, laminated microvia substrates and 3-D folded flex stacks with multiple chips and embedded passives will also be an attractive design point.

Until the Year 2001, SoC has been the approach of choice for many digital applications because the extant semiconductor technologies have supported it. Chip sizes have been below 20 mm square, performances have been below 1.5 GHz, yield and process complexity have been acceptable and packages with sufficient TFW wiring have been too expensive. In the future, however, this situation should change to favor SiP.

All of this points to SiP becoming more important with time. This will naturally occur if system engineers look at differential costs and overall program risk at the product or box level. Whenever this occurs, all interactive E-packaging costs are taken in to account. This means that cost/performance factors associated with chips, modules, PWBs, connectors and cooling should all be globally optimized. Because of competitive pressures, the simpler technique used by developers to bargain for a lower price on each part independently will cease to be adequate. When the designer starts looking at the big picture; sub-optimization will cease, design points will change and technological advances like SiP will become more important.

NOTES

1   The 2001 ITRS summary document can be found at:
    "http://public.itrs.net/Files/2001ITRS/ExecSum.pdf".
2   See the 1999 ITRS at "http://public.itrs.net/#WhatsOn".
3   It should be mentioned that the ITRS is not a prediction document; rather, it defines the Grand Challenges required to maintain the historical trends of Moore's Law which predicts a doubling in chip function every 18 to 24 months. If any of the Grand Challenges (planned inventions) do not occur on time, the predicted design points will either occur later or not at all. Beginning in 2001, the author has noted that the semiconductor leaders are beginning to admit that the straightforward device scaling days are nearing an end and the planned invention era is now upon them.
4   In the Systems Drivers Chapter of the 2001 ITRS, there is an implication that the main definition of an SoC is a chip consisting of many reusable cores of existing functions or IP (Intellectual Property from many sources) rather than the earlier implication that it will be a single chip system.
5   See Assembly & Packaging at: "http://public.itrs.net/Files/2001ITRS/Home.htm".
6   Total pins per chip and total I/O per SCM for a single chip on an SCM differ because a large percentage of the power and ground chip pins are combined within the SCM. In the typical case, the SCM substrate pin count is lower because there are fewer power pins. Note, however, that in the later years chip and package I/O become the same in the ITRS. This is the case because the signal to power pin ratio for both chips and packages are assumed to be 1:1 to support high currents and high frequency signal shielding requirements.
7   The OLGA technology is based upon an organic technology originally called SLC (Surface Laminar Circuit) when developed by IBM's facility in Yasu, Japan in 1989. [3]
8   As an example, beginning with the Athlon XP series of microprocessors in 2001, AMD has switched to organic built-up substrates.
9   The initial McKinley (2002) chip has been built at the lower risk 0.18 µm process node and its area has been reported to be 464 mm2 at 1 Ghz. clock frequency. Using a mature process technology first is normally done to demonstrate functionality but the chip is too large for a profitable business case. The production version will be done in 0.13 µm technology in 2003 at approximately half the chip area. The McKinley chip contains 221 million transistors.
10 In ITRS parlance, a "node" refers to a semiconductor ground rule generation; e.g., the 0.18 mm and the 0.13 mm nodes are successive processes that use 0.18 and 0.13 nm lithographic exposure equipment, respectively.
11 In any system with multi-processors and multi-caches, a means has to be provided to know where the most recent or "good" data is located. If write through of data to a central cache with a directory is used such as the L2.5 herein, this is where the good data is located. This is referred to as a coherent cache system and it avoids the long latencies associated with "snooping" all the non-coherent caches to find the most recent instantiation of the data when a central cache is not used. The most powerful servers typically use the cache coherent design with a central cache.

12 Silver has a slightly better conductivity than copper but it is rarely used as a printed wire because it is more expensive and it is more susceptible to oxidation.

13 To glean this information from Figure 4, multiply the power density values by 3.1 (cm^2), the area of the chip.

14 TFT-J is built up layer-by-layer on a large glass panel. It is then joined to either chip studs or surface pads on a multi-layer packaging substrate and the glass panel is released using a laser technique.[10]

REFERENCES

1. E. Davidson, *et. al.,* "Long Lossy Lines (L^3) and Their Impact Upon Large Chip Performance," IEEE Trans-CPMT-B, pp. 361-374, Vol. 20, No. 4 (1997).

2. L. Schaper and T. Lenihan, "Passives go Into Hiding,"*Advanced Packaging Magazine*, pp. 22-26, Feb. 1998.

3. Y. Tsukada, *et. al.,* "Surface Laminar Circuit and Flip Chip Packaging," *Proc. 42nd ECTC Conference,* pp. 22-27, 1992.

4. R. Shukla, *et. al.,* "Flip Chip CPU Package Technology at Intel: a Technology and Manufacturing Overview," *Proc. 49th Electronic Components and Technology Conf.,* pp. 945-949, May 1999.

5. E. Davidson, "SoC or SoP? A Balanced Approach!," *51st Electronic Components and Technology Conf.,* pp. 529-535, May 2001.

6. A. Blodgett, "A Multi-layer Ceramic Multi-chip Module," *IEEE Trans. Components, Hybrids, Manuf. Technol.,* CHMT-3, May 1980.

7. G. Dudeck and J. Dudeck, "Design Considerations and Packaging of a Pentium Pro Processor Based Multi-chip Module for High Performance Workstations and Servers," *1998 IEEE Symposium on IC/Package Design Integration,* pp. 9-15, Feb. 1998.

8. G. Katopis, *et. al.,* "MCM Technology and Design for the S/390 G5 System,"*IBM J. of Res. and Develop.,* pp. 621-650, Vol. 43, No. 5/6 (1999).

9. E. Davidson, "Large Chip vs. MCM for a High Performance System,"*IEEE Micro,* pp. 33-41, Vol. 18 No. 4, July/Aug. 1998.

10. H. Braunisch, et. al., "Electrical Performance of Bumpless Build-up Layer Packaging,"*52nd Electronic Components and Technology Conf., May, 2002.*

11. C. Prasad, et. al., "Multi-level Thin-film Electronic Packaging Structure and Related Method," U.S. Patent 6,281,452, Aug. 2001.

12. H. B. Pogge, et al, "Bridging the Chip/Package Process Divide," Proc.of Adv. Metallization Conf., Montreal, Oct. 2001.

Chapter 3

PACKAGING TECHNOLOGIES FOR FLEXIBLE SYSTEMS

Christine Kallmayer
Fraunhofer IZM, Gustav-Meyer-Allee 25, 13355 Berlin, Germany, www.izm.fraunhofer.de

1. INTRODUCTION

The overall drive to smaller and lighter packages together with a continuously increasing I/O count has lead to an increase in interest in flexible substrates for electronic packaging. With the growing number of applications for flexible modules the variety of requirements for these packages increased as well. Different technologies had to be developed to meet the needs of products ranging from CSP´s over Smart Cards to complex systems e.g. medical implants. Different soldering techniques as well as adhesive joining processes have been optimized and qualified for flexible substrates. Simultaneously the development of thin flexible dies provided a key technology for flexible systems.

The assembly processes for thin IC´s on thin flexible substrates require thin interconnections. These require new or modified bonding processes. Their behavior regarding mechanical stress as well as thermal ageing is different from conventional contacts and is still under study.

For portable consumer applications as well as for products with a high need for miniaturization (e.g. medical implants) flexible substrates allow a 3-dimensional shape of the electronic circuit which can be adapted to the individual product requirements.

For the realization of systems in package (SIP) this concept automatically leads to the more ambitious development of flex-based stackable CSP´s and folded 3D modules. This new type of packages finally makes use of all advantages of flexible substrates:

- Mechanical flexibility
- Minimum weight
- Minimum volume
- High density
- High reliability

2.   TECHNOLOGICAL PREREQUISITES

The capability to thin IC´s down to less than 50 μm is the enabling technology for flat and flexible systems. These dies show a high mechanical flexibility comparable to polymeric foils and are therefore ideally suited for assembly on polymer tapes. Their availability will make it possible to produce low-cost products like smart labels on polymers or on paper in reel to reel processes[i]. On the other hand it provides access to 3D packaging technologies either by embedding the dies, stacking thin packages or by folding the assembled flexible modules.

Besides the dies, the substrate and its mechanical properties are more important in such advanced packages. If bending and even folding of the substrate is required, the adhesion of the conductor lines and their ductility have to be optimized beyond conventional requirements. Best results are obtained if the conductor lines are embedded in the two layers of the same polymer thus being in the neutral zone for bending [ii]. The availability of flexible circuit boards with structures down to 20 μm lines and space is important to allow flip chip assembly while eventually reducing the number of layers. Less layers not only reduce cost but also help to maintain the flexibility of the module.

3.   FLIP CHIP TECHNOLOGIES

For the assembly of bare dies on flexible circuit boards an number of flip chip technologies is available. Generally the conventional flip chip processes can be applied also for flexible substrates capable of withstanding the process temperatures and pressures. The major modification is the handling of the flex during the process. But there are also approaches to flip chip assembly which lead only to good results in combination with flex substrates. If low cost polymers are used the processes have to be modified in order to reduce thermal load. This is done by developing processes with lower peak temperatures and shorter cycle times.

3.1   Thermode bonding

Using very small solder volumes deposited by immersion soldering together with extremely fast heating ramps (pulse heating) during soldering has been shown to be a promising technology for flip chip assembly on flex down to finest pitches of 40 μm and also for high melting solders [iii]. These thin interconnections are an important step forward regarding thin modules as well as 3-dimensional assemblies.

Due to the thin solder layer between chip metallization and conductor line the understanding of metallurgical interactions and the characteristics of dominating intermetallic compounds has become even more important than for conventional flip chip assemblies. Controlling and understanding the impact of the intermetallic phases is already a key parameter for the optimization of the bonding process. The ageing behavior and the mechanical stability of such contacts are currently still under investigation [iv].

3.1.1   Bumping

For low-cost wafer bumping stencil printing of solder paste on electroless Ni/Au Under Bump Metalliziation (UBM) is a mature process and already widely used. A large variety of solder pastes including lead-free alloys are available. However due to available solder pastes and stencil geometries this process is still limited in pitch down to 200 μm for high volumes and 100 μm for special layouts and laboratory scale experiments.

For very fine pitches down to 40 μm electroplating has been the only wafer level solution but with new developments immersion solder bumping (ISB) can be an extremely flexible low-cost alternative. Due to its good wettability Ni/Au is also used as UBM for this process. In the second step the wafer is immersed in liquid solder as shown in figure #3-1. The Ni UBM is wetted and a small solder cap is formed on top of Ni.

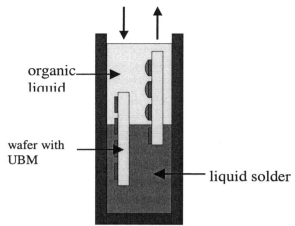

Figure #3-1. Principle of immersion solder bumping

An organic liquid prevents oxidation of solder and improves wettability. After soldering residues can be easily removed. The process has been tested successfully for wafers up to 6 inch [v],[vi].

Figure #3-2 gives an overview Ni/Au bumps with 5 μm thickness immersed with eutectic PbSn solder at 50 μm pitch. The solder cap height strongly depends on the pad size, Ni bump height and the surface tension of the solder material. Besides eutectic PbSn the ISB technology has also been demonstrated with various leadfree solder alloys including SnAg and eutectic AuSn. For the different alloys only the process temperature has to be changed according to the liquidus temperature of the alloy.

The average solder cap height for eutectic PbSn at a pad diameter of 50 μm is 8 μm, at a pad diameter of 20 μm only 3 μm of solder are deposited. The deviation of the bump height is significantly larger than for stencil printed or electroplated bumps but still suitable for flip chip bonding on flex as long as pressure is applied during soldering.

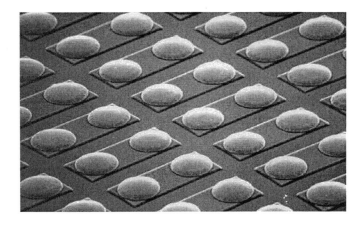

Figure #3-2. Overview of immersed Ni/Au bumps at 50 μm pitch

3.1.2   Bonding technology

Conventional flip chip assembly by reflow soldering is not suitable for those small solder volumes with the typical large deviation. In addition the gap height is reduced so that the flow of normal underfiller will lead to very long cycle times if it is possible at all. To prevent these problems the preapplication of "noflow" underfiller [vii] is a promising approach. Stencil printing, dispensing or dipping are suited for this process depending on the desired volume and the required accuracy. In order to prevent a contamination of the bonding tool during the thermode bonding process it is necessary to achieve a good volume and positioning control of the applied underfiller depot. In figure #3-3 the printing method is depicted. Subsequently a thermode bonding process is performed during which the solder joints are formed and the curing of the underfiller is initiated.

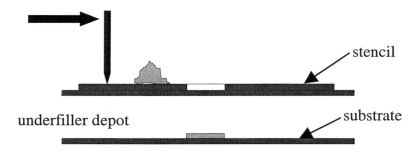

Figure #3-3. Printing of "noflow" underfiller

Thermode bonding processes are characterized by using pulse heated thermodes for high heating rates and fast cycle times [viii]. The bonders suitable for this process are either modified TAB bonders or – for higher accuracy ($< 5\mu$m) – flip chip bonders typically used for adhesive joining.

In figure #3-4 the most important steps for the thermode bonding process using "noflow" underfiller are shown. As the solder volume of each bump is very small the problem of solder bridges between adjacent contacts is negligible. Therefore it is possible to use a force controlled bonding profile instead of height control. During the soldering process the oxidized surface of the immersion solder bumps has to be cracked. Typically a bonding force of 0,2 N/bump at 100 μm pitch leads to good results. The bonding temperature depends on the used solder alloy but has to be significantly higher than the melting point. In order to reduce the phase formation between substrate metallization, UBM and solder the time above the liquidus temperature has to remain as short as possible. On the other hand the underfiller has to start curing to tack the chip onto the substrate surface during the soldering process. This is necessary because of the lower mechanical stability of these small and thin soldered contacts in comparison to conventional higher soldered bumps. The fast process provides selective heating of the interconnection areas, thereby avoiding softening and damaging of the flex even if high melting solders like eutectic AuSn are used.

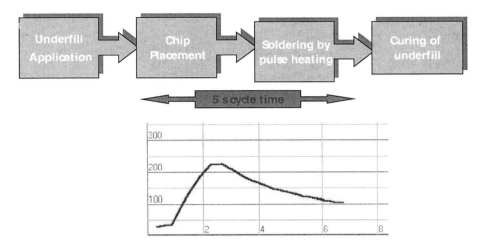

Figure #3-4. Thermode bonding process

Thermode bonding offers a large process window. In the parameter set only the bonding force depends on the number of I/Os of the die. The resulting flip chip contacts are characterized by excellent fillet formation and a very low stand off height of less than 10 μm as shown in figure #3-5 a and b. The suitability of this technology for ultra fine pitch flip chip assembly is shown in figure #3-6 which shows a cross section of a 40 μm pitch test vehicle mounted on flex with eutectic AuSn solder. It is obvious from these cross sections that the major difference to conventional large flip chip contacts is the influence of the intermetallic phase which form the largest part of the contact volume. As data on their material properties are not yet available the prediction of the behavior of such structures is difficult. First reliability studies suggest that thin PbSn solder joints are only suited for limited reliability requirements but eutectic AuSn is highly reliable in humidity as well as high temperatures and thermal cycles.

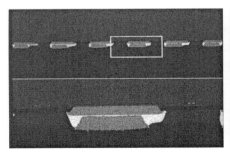

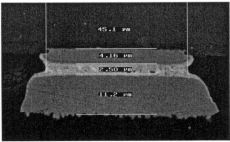

a)   b)

Figure #3-5. Typical flip chip contacts otained by thermode bonding
a) ISB eutectic AuSn solder, b) ISB eutectic PbSn solder)

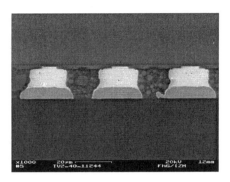

Figure #3-6. Flip chip solder joints at 40 μm pitch with electroplated AuSn solder

3.2   Adhesive bonding

Adhesive bonding using anisotropic conductive adhesive (ACA) is a well known and successful technology for the assembly of fine pitch driver ICs on glass for LCD applications. Recently ACA and also NCA (nonconductive adhesive) bonding is becoming the most promising technology for mounting thin IC´s on flexible substrates e.g. for smart labels is flip chip bonding. The process requires a pad metallization on the die typically of electroplated Au, electroless Ni/Au or electroless Pd. While NCA bonding shows optimum results using bumps with a topography (Au stud bumps, electroless Pd) the ACA bonding has been demonstrated for a wider variety of bump types.

The deposition of the adhesive can be performed by dispensing, stencil printing or dipping depending on the positioning accuracy and volume control required. Typically dispensing the adhesive without a special pattern will lead to good results but for chips thinner than 50 μm either special dispensers for ultra small volumes or stencil printing have to be chosen.

Bonding is performed using a flip chip bonder holding pressure and temperature over the curing time of the adhesive (5-20s) [ix]. These short cycle times allow the use of adhesive bonding for reel to reel assembly of low cost products. For ACA the process provides the electrical contact by clamping the conductive filler particles of the adhesive between bumps and conductor line. The resulting contact resistance for 100 μm contacts is ≤ 20 mΩ as there are only few particles on the contact area. A typical contact with

Ni/Au bumps and Au filler particles is shown in figure #3-7. The application is shown in .The NCA process provides a metallic contact directly between bump and conductor line which leads to a lower contact resistance of ≤ 10 mΩ for the same pad size.

On flexible substrates these processes allow a low stand off together with very good reliability for ambient and operating temperatures below the glass transistion temperature of the adhesive.

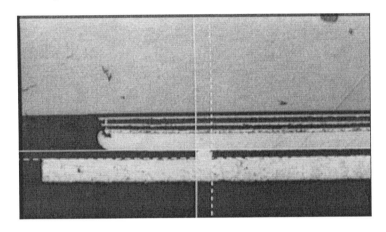

Figure #3-7. Typical ACA flip chip contact on 10 μm Polyimid flex

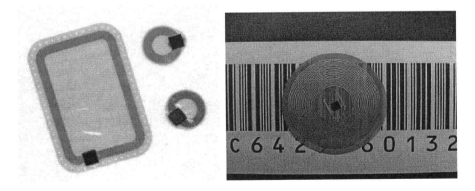

Figure #3-8. Passive transponders in ACA technology on thin flexible Polyimid substrates

3.3   Reflow soldering

Flip chip assembly using Pb/Sn eutectic solder and reflow soldering is the conventional approach to obtain the advantages of this miniaturized packaging method[x]. The dies are typically bumped by electroplating or stencil printing on electroless Ni/Au on wafer level. The singulated dies are then placed on the printed circuit board and soldered in a conveyor oven. For this process a good infrastructure already exists regarding bumping service, materials (fluxes, pastes, underfillers) and equipment or manufacturers.

Due to the upcoming ban of lead in solders flip chip soldering with leadfree solders has been investigated in the last years with good progress. Polyimide based flexible circuit boards have been used commercially with eutectic PbSn but are well suited for leadfree solders with their higher melting points. This combination will be interesting for future applications with high reliability requirements. For Multi Chip Modules (MCMs) and System-in Package (SIP) the compatibility with the assembly of the necessary SMD components is the deciding advantage over the competing processes. Two examples from areas where reliability is required to insure safety are shown in figure#3-9: a flexible airbag controller module and a flexible module for ultra small hearing aids.

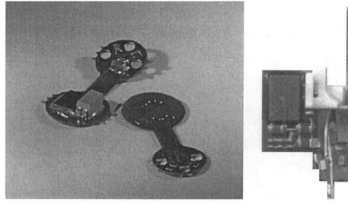

Figure #3-9. Flexible modules in flip chip technology

4.   3D-PACKAGING CONCEPTS

The assembly technologies presented provide the minimum package volume available today and enable highly integrated subsystems, so called Systems-in-Package (SIP). For some applications this degree of miniaturization still is not sufficient. Only the application of 3-D-packaging methods can provide further decrease in footprint and volume together with a flexibility in shape not available in todays electronic modules. The advanced flip chip technologies combined with thin chips which were presented are the prerequisites for 3-D solutions. There are three basically different approaches for 3D modules with interposers:

- foldable modules
- stackable modules
- direct integration of ICs in flex or PCB.

Additionally there is the possibility to stack thinned and planarized wafers and create contacts between the layers by etching throughholes and metallizing them. This substrateless technology will not be described here in detail.

4.1   Folded packages

Folded flexible packages are especially interesting for memory modules, medical products (e.g. implants) and miniaturized microsystems (e.g. smart dust). The first demands for such a technology resulted from the necessity to integrate a constantly increasing amount of memory in smaller and smaller products. As the memory dies also increased in size the only way to obtain reasonably small footprints for memory packages is 3D assembly. Thin Polyimide substrates e.g. by Hightec, Switzerland have been used to create this type of folded package typically with one IC type[ii]. The assembly can be perfomed with any flip chip technology. As a low stand off is an advantage adhesive bonding and thermode bonding with ISB bumps are recommended. During the bonding process the thin substrates are ideally mounted on a carrier. If the flex has been structured in thin film technology it can conveniently remain on the carrier used during processing. The interconnection with the board is typically achieved by an array of BGA or CSP solder balls. In figure #3-10 an example for a folded structure is shown.

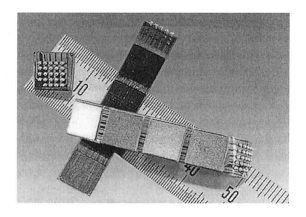

Figure #3-10. Folded package in Flip Chip on flex technology (FCOF) by Hightec SA

For 3 D realization of SIP the concept in figure #3-11 can be more suited as most passive components sensors and microsystem components are not thin enough for the completely flexible solution. Here a rigid flex substrate is assembled with all components and then folded in the final shape. The assembly can therefore be realized by conventional flip chip and SMD technology.

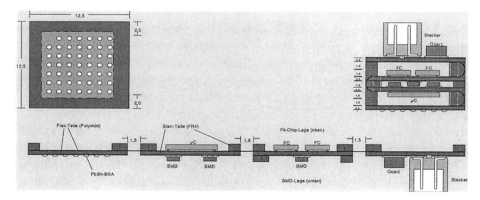

Figure #3-11. Concept for folded SIP on rigid flex

4.2 Stacked packages

Stackable modules are used as a matchbox approach to flexible products including sensors and actuators. Different concepts are followed regarding interposer materials and assembly technologies. Stackable packages are available in flip-chip and wirebond technology on ceramic, FR4 and flex substrates depending on the area of application.

The StackPack is a logical development towards high integration, low volume and high functionality solutions e.g. used in Mechatronics. This package is based on the combination of a novel packaging solution and the high integration features of FlexPack[xi] , a chip size package developed at IZM. StackPack offers also a high rate of manufacturability as it is possible to manufacture even Multi Chip Packages, including discrete components using a reel to reel processes. StackPack processing starts with a polyimide tape with a photostructured metallization, as depicted in figure #3-12. The footprint to the bottom package or substrate and the package on top is standardized, allowing fast redesigns by changing one single interconnection layer. The IC's and discrete components are assembled using fast and reliable laser bonding (FPC) or thermode bonding[xi].

The flexible tape is no longer just the carrier for the components, it is an an integral part of the whole package, providing shape and stiffness. The module is then molded with a thermosetting resin with high heat conductivity, to achieve good thermal performance. The material system chosen offers all advanced properties for a package used under harsh environmental conditions. After the molding-process the modules are separated and the top layer is folded and fixed to the top of the package (see figure #3-12). The result is a package of very small size and low volume, which is assembled in low cost standard IC assembly processes. This module fulfils almost all requirements due to the above definition.

Using StackPack it is possible to mount different types of modules in stacks together and connecting it to a standard bus system as shown in the Mechatronic solution using ceramic TB-BGA's[xii]. A package assembled like StackPack, including the sensor and housing of the sensor allows also full testability of the assembled device and eliminates several interconnection layers. This leads to an overall reduction of interconnections of approximately 35-40 % providing a drastic increase in package reliability.

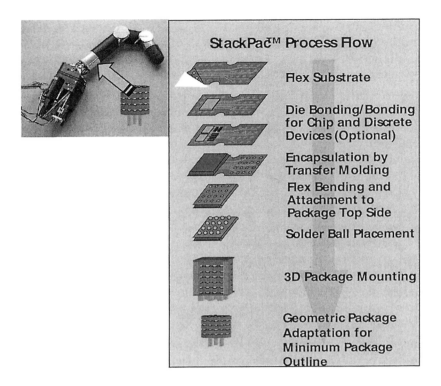

Figure #3-12. Stackable flex-based CSP "Stack PAC"

4.3 Chip in Polymer (CiP)

3D integration can also be realized by integrating the thinned IC´s directly in the circuit board (flexible or rigid) itself serving as a miniaturized sub-system on a motherboard. This new technology is currently investigated and is suitable e.g. for mobile phones.

The underlying technology of these approaches described in [xiii] and [xiv] lends itself easily to embedding technologies. Such embedding technologies have been investigated and described previously [xv, xvi, xvii] but have been lacking the technological advancement available today. Thin chips facilitate the use and applicability of the large area manufacturing techniques in the PCB industry.

Two different approaches are possible for the integration of dies in circuit boards. One uses liquid polymers for covering the die on the core layer the other uses conventional resin clad copper foils (RCC) suitable for microvia PCB manufacturing.

4.3.1   Chip Assembly

For chip assembly to the substrate, the die must be glued to the substrate surface with a suitable adhesive and placement equipment. The mass manufacturability of thin chip assembly has been shown by e.g. Toshiba and Sharp for their multi chip stacked CSP. As in these CSPs wirebonds are used initial placement accuracy is not very critical. In contrast to this, placement is a very important issue for the embedding technique as simultaneous exposure by a predefined pattern is used for the structuring of vias and metal routing. Therefore vision systems with top-down camera/optics and respective alignment patterns must be incorporated into the placement machine to provide an accuracy of ~15μm. Also the die attach adhesive must allow to glue the chip with minimal standoff (figure #3-9), as too much standoff will prevent successful subsequent dielectric coating.

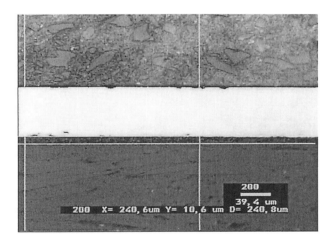

Figure #3-13. Thin die with thin die attach adhesive

4.3.2   Dielectric Coating

Covering the mounted IC with the subsequent dielectric layer for interconnect and additional routing requires the deposition of a suitable material. The use of photosensitive polyimide and BCB was successfully shown previously [xviii]. As neither processing temperatures or material are compatible to PCB material (e.g. FR4, FR5, CE etc.), epoxy materials were selected to serve as a dielectric. Although RCC foils lend themselves easily to lamination and laser drilling, a spin on PCB compatible epoxy resin was used for the layer deposition as a first step.

4.3.3 Via Formation & Metal Deposition

In today´s high density PCB manufacturing, laser drilling with either excimer lasers or CO2 lasers as well as plasma etching is a well established process for in-dielectric via formation. Photo definition is another well known process, however it allows only limited material selection. When a photo sensitive epoxy material is selected, transparent films or glass masks can be used to structure the vias through the dielectric to the contacts of the chip. In our case, a photosensitive dielectric was spun on a 100x100mm carrier with embedded chips. Using a glass mask and a collimated UV light source the vias were formed.

Typically, in a PCB manufacturing facility, after via formation an electroless Cu seed layer is applied and reinforced by electro-deposited 5-15um Cu. For ease of processing in the CiP-process 6um of electroless Cu, as described in [xix], are deposited. Finally, the subsequent layers can be applied and structured as needed.

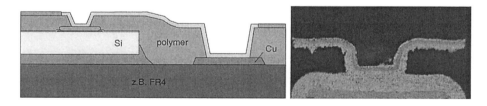

a)   b)

Figure #3-14. a)Schematic of the embedded structure, b) cross section of chip contact via

4.3.4 Thermal Management

The assembly of the thin chip to the substrate may be accomplished either on the base substrate itself or on a copper layer (e.g. large area circuitry) underneath (figure #3- 6). The latter approach mimicks a built-in heat spreader, that can be attached by thermal vias or by heat traps to a larger heat sink.

If for any reason this approach is not sufficient, active cooling by heat pipes or direct liquid cooling has been demonstrated previously to be incorporated into the PCB manufacturing process.

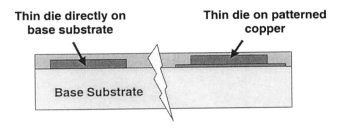

Figure #3-15. Different methods for chip attach to the substrat

4.3.5 Applications

The CiP technology is suitable for the production of smart circuit boards which contain active and passive components. But it can also be applied for mass lamination capable multilayer multi chip stackable CSPs providing ultimate flexibility and miniaturization. A schematic of such a stackable module is shown in figure #3-16.

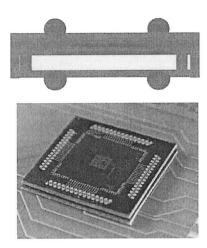

Figure #3-16. Schematic structure of a multilayer UT-CSP, demonstrator example of stacked UT-CSP

5.   CONCLUSIONS

The availability of thin IC´s together with an number of adequate technologies for ultra-thin assembly provide access to a whole new range of innovative products and applications. The possibilities range from smart labels to 3-dimensional high performance systems including microelectronic and microsystem components. Folding, stacking and integration in boards will lead to a breakthrough in miniaturized mobile electronic products, making them smarter and smaller.

[i] R. Aschenbrenner et al. , „Concepts for Ultra Thin Packaging Technologies" Conference Adhesives in Electronics 2000, Espoo, Finland, June 18-21, 2000,

[ii] Alexander Fach et al., "Multilayer polyimide film substrates for interconnections in Microsystems", Proc. Microsystem Technologies 5, 1999

[iii] Barbara Pahl et al., „A Thermode Bonding Process for Fine Pitch Flip Chip Applications Down to 40 Micron", Proc. EMAP, Korea, 2001

[iv] Barbara Pahl et al., A Thermode Bonding Process for Fine Pitch Flip Chip Applications on Flexible Substrates, to be published Proc. IMAPS Nordic Conference Oct. 2002, Stockholm, Sweden

[v] Patent Nr.WO96/08337, "Method of applying solder to connection surfaces, and method of producing a solder alloy"

[vi] Patent Nr. DE 44 32 774 C, "Verfahren zur Herstellung meniskusförmiger Lotbumps"

[vii] Christine Kallmayer et al., Processing Design Rules for Reliable Reflowable Underfill Application, Proc. 51st Electronic Components and Packaging Conference, Orlando, May 2001

[viii] Christine Kallmayer et al., "Chip interconnection technologies on flexible circuits", Proc. of IMAPS Flip Chip Workshop, March 2000,Braselton, USA

[ix] Rolf Aschenbrenner *et al*, "Evaluation of Adhesive Flip Chip Bonding on Flexible Substrates", *Proc. Flexcon'97*, Sunnyvale, 1997

[x] Erik Jung et al., "Experience with a fully automatic flip chip assembly line integrating SMT". Proc. NEPCON West, March 1998, Anaheim, USA

[xi] Christine Kallmayer, "A new Approach to Chip Size Package Using Meniscus Soldeirng and FPC Bonding", Proc. 47th Electronic Components and Technology Conference, ECTC '97, San Jose, California, 1997

[xii] Hr. Reichl, V. Grosser, "Overview and development trends in the field of MEMS packaging", IEEE MEMS 2001, Interlaken, 2001

[xiii] S. Andersson, "Packaging - Providing Solutions between Silicon and System", IEMT 2000, Santa Clara, Oct. 2000, Keynote Talk

[xiv] M. Schünemann et al., "MEMS Modular Packaging and Interfaces", 50th IEEE Electronic Components and Technology Conference, Las Vegas, May 2000

[xv] Eichelberger et al., US Patent US5841193: "Single chip modules, repairable multichip modules, and methods of fabrication thereof"

[xvi] Hahn et al., "Fabrication of high power MCM by planar embedding technique and active cooling", Micro Systems Technology, 1996

[xvii] Klose: "Chip First Systeme und –Gehäuse", Thesis from Univ. Siegen, Germany, Jan. 2000, ISBN 3-8265-6952-0

[xviii] Hahn et al., "Fabrication of high power MCM by planar embedding technique and active cooling", Micro Systems Technology, 1996

[xix] Ostmann et al, " Development of an Electroless Redistribution Process", Proc. IMAPS Europe 1999, Harrowgate, June 1999

Chapter 4

LOW-PROFILE AND FLEXIBLE ELECTRONIC ASSEMBLIES USING ULTRA-THIN SILICON – THE EUROPEAN FLEX-SI PROJECT

Thomas Harder, Wolfgang Reinert
Fraunhofer Institute for Silicon Technology (ISIT)
Itzehoe, Germany

1   INTRODUCTION

Today there is a strong trend towards thin and flexible packages which is driven by different industrial applications e.g. smart cards and electronic labels. Furthermore, several technical approaches for ultra-thin chip stacking leading to 3D assemblies are followed. In the European FLEX-SI Project (IST-99-10205) the eight partners Philips Semiconductors (A), W.S.I. (F), Datacon (A), Nokia (FIN), Oticon (DK), PAV CARD (D), Helsinki University of Technology (FIN) and Fraunhofer ISIT (D) are working together on an industrial approach of low-profile packaging with ultra-thin silicon chips.

The FLEX-SI project aims at the development of different thin electronic packaging solutions based on ultra-thin flexible Si chips with a thickness < 50 μm. Therefore, industrial processes for the wafer thinning, handling and mounting/taping as well as shipping techniques have been developed. Furthermore, the materials properties of ultra-thin silicon especially the electrical performance of circuits under bending stress are investigated including the analysis of long term behaviour and reliability.

Derived from different industrial applications three ultra-thin packaging technologies are realized: multi chip module with active component integration into flexible substrates (MCM-L), chip-on-chip and chip-in-flex. The electrical interconnection is performed mainly by a small gap, flip chip technology. Demonstrators from different application fields employing these ultra-thin packaging technologies are assembled and qualified:

- Mobile telecommunication device employing MCM-L

- Hearing aid using chip-on-chip

- Identification system (smart label) with chip-in-foil

The main objective of the project is to provide industrial processes enabling ultra-thin packaging solutions and flexible electronic assemblies which will on the one side improve existing products e.g. in mobile telecommunication and medical applications and on the other side will enable totally new products with concealed electronics.

Key issues

- Wafer thinning down to a rest thickness < 50 μm including the necessary stress release

- Material properties of ultra-thin silicon including the electronic performance of circuits under bending stress, long-term behaviour and reliability; improved test methods for thin wafers

- Industrial handling and shipping techniques for ultra-thin wafers; proto-type equipment optimized to handle thin silicon e.g. flip chip die bonder (Datacon)

- High density packaging technologies leading to paper-thin assemblies

The ultra-thin wafers of 50 μm rest thickness were provided by W.S.I., a service company specialized in thinning wafers. The W.S.I. thinning technique which is also working with wafers with ink dots or bumps has the main advantage of complete de-stressing leading to flexible and mechanically robust wafers.

4.2 ELECTRICAL PERFORMANCE OF ULTRA-THIN WAFERS UNDER BENDING STRESS

The characterisation of electronic devices under mechanical stress is one of the objectives in the FLEX-SI project. Functionality of IC's under mechanical stress is important for all kinds of chips mounted on a flexible substrate, like assemblies on flexible PCB's or Smart Labels. In conventional assemblies, but also in smart card applications with thin dice of about 150 μm thickness, the silicon bulk material can take up the mechanical forces. As die thickness decreases, the circuitry on the chip surface is more and more exposed to deformation.

Therefore, the main intention of this work is to show, if electronic circuits on a silicon IC can still be functional under static mechanical stress.

4.2.1   Experimental

For this investigation a special test wafer was designed, see figure 1. The mask contains:

- a standard PCM (process control module) field
- several test-devices (diodes, resistors, transistors)
- EEPROM cells
- several versions of the I•Code IC for rf identification systems

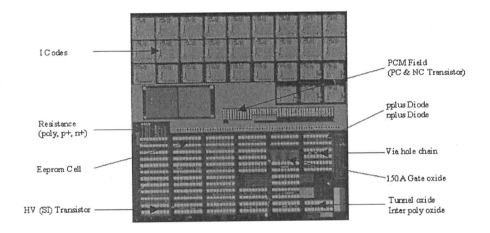

Fig. 1: Design of the CMOS test wafer.

Since the effect of mechanical deformation should be tested, it was decided to compare results measured on ultra-thin test devices in flat and bent state, respectively. The overall test-flow is depicted in figure 2.

A batch of test-wafers was split in two parts. One part (#1) passed an extended automatic PCM test in the Philips testcenter, followed by wafer thinning to 50 μm at W.S.I. . After that, the thin wafers were attached to rigid carriers (i.e. silicon dummy wafers) and a 2nd PCM test was performed. Attachment of the thin wafers to a substrate was done by laser drilling 100 holes into each carrier and mounting the thin wafers on the carriers on a vacuum chuck using a special tape with thermal release (Revalpha tape from Nitto).

For the other part (#2) the procedure was different: A dedicated set of characteristics exceeding the standard fab-tests was recorded manually. After thinning, the wafers were diced into single reticles. Reticles from different positions on the wafer were selected for measurement. Since the test should reveal the properties of mechanically stressed devices, static tensile stress was applied to the reticles by attaching them to a cylindrical carrier with a double–sided adhesive tape. It was decided to chose two different cylinder diameters to be able to compare results from reticles, which have been stressed to a different extent. Cylinder diameters of 50 mm and 70 mm were chosen.

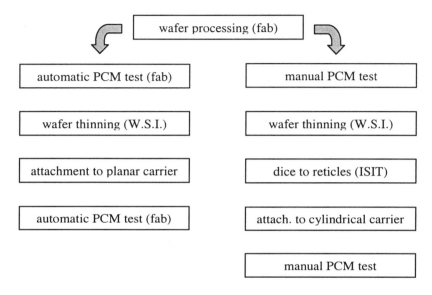

Fig. 2: Test flow to investigate the effects of wafer thinning and static mechanical stress on the performance of electronic devices.

In order to be able to find different effects for different bending directions, half the samples were bent in x-direction (0°) and the other half in y-direction (90°). So finally the test plan ends up with four measurements for each test device:

1. sample on 50 mm – carrier; orientation 0° (d50 0°)
2. sample on 50 mm – carrier; orientation 90° (d50 90°)
3. sample on 70 mm – carrier; orientation 0° (d70 0°)
4. sample on 70 mm – carrier; orientation 90° (d70 90°)

All measurements were performed using a Hewlett Packard semiconductor analyser (HP 4155A and HP 41501A). In figure 3 the four cylindrical carriers can be seen. Every carrier holds at least three samples, which are mounted with double-sided adhesive tape. Also the perpendicular bending directions on cylinders on the left- and right hand side can be seen in the picture.
These cylindrical carriers with the test samples were placed on a probestation and all measurements were done manually.

d50 0°

d50 90°

d70 90°

d70 0°

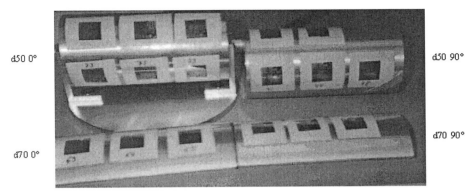

Fig. 3: Cylindrical carriers with samples attached.

4.2.2 Results

The aim of this work was to find out, if electronic circuits can still be functional under strong mechanical stress. Indeed the strain on the samples was very high in these tests and most of the samples on the 50 mm − cylinders cracked when left overnight. So, for devices with 50 μm rest thickness the applied tensile stress pretty much represents an upper limit.

The most important result in this context is that all test devices were still operating after bending. There are, however, several effects resulting from mechanical stress, which are worth being discussed.

All field effect transistors tested in this procedure exhibited similar effects as far as drain currents are concerned. There seems to be a reproducible influence of bending stress on the transistor characteristics. Moreover, also the bending direction makes a significant difference. The effects are illustrated in figure 4.

Note that the device characteristics for different bending directions (0° and 90°) were not measured on the same samples, but on samples taken from different wafers. Therefore, the curves cannot be compared to each other directly. We can only compare the data obtained from the thick wafer and the corresponding thin and bent wafer (i.e. we cannot compare d70 0° and d70 90° but only d70 0° and the corresponding thick wafer etc.).

n-channel transistor 10/0.8

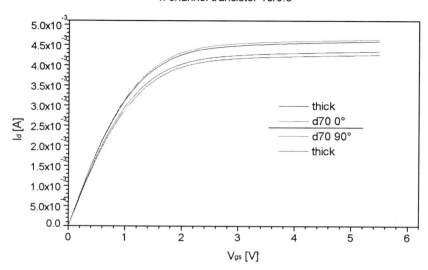

p-channel transistor 10/0.8

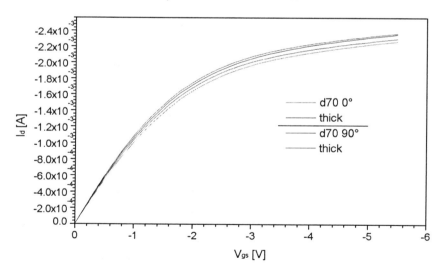

Fig. 4: Characteristics of n-channel and p-channel MOSFETs under static mechanical stress.

The following results are shown in figure 4:

- *p-channel transistor:* drain current is slightly lower for devices bent in 0° - direction compared to thick wafer; for bending in 90° - direction, the current is higher!
- *n-channel transistor:* reverse effect; here the current is higher for bending in 0° - direction and lower in 90° - direction. The difference is smaller compared to PMOS.

In both cases, the same effects have been found for the 50 mm – substrates and also the magnitude of the effect was almost the same. In table 1 the results for all MOSFETs are summarized.

type	bending parameters	current change [%]
PC 10/0.8	d70 0°	-3.8
	d70 90°	+3.0
	d50 0°	-3.9
	d50 90°	+1.7
NC 10/0.8	d70 0°	+0.9
	d70 90°	-1.9
	d50 0°	+0.5
	d50 90°	-1.9

Table 1: Change in the drain current of MOSFETS at bending in two perpendicular directions.

The question arises, what exactly causes the changes in the transistor characteristics. It could be either thinning or the mechanical stress or both. To clarify this issue, a set of wafers (part #1) was automatically measured before and after thinning. Results for the saturation current of both types of FETs in flat, non-bent state are shown in table 2. From table 1 and 2 it can be seen that actually both, thinning and bending, contribute to the change of the characteristics. However, the effect of the mechanical stress is dominating and depends on the bending direction.

Figure 5 shows both contributions, the data were measured on the very same sample.

type	current change [%]
PC 10/0.8	-0.9
NC 10/0.8	-0.5

Table 2: Comparison of saturation currents for thick and thin wafers.

PC transistor 10/0.8

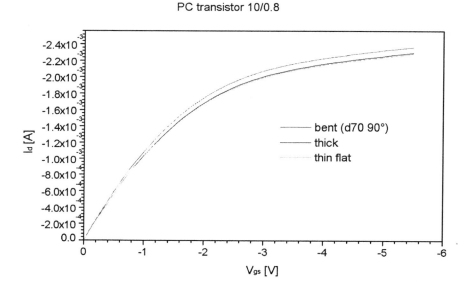

Fig. 5: Comparison of the drain current of a p-channel MOSFET after thinning and after thinning and bending.

4.2.3 Discussion and Outlook

The first – and most important – result is that almost all devices have still been operational after bending the samples, even though the stress applied was quite high and the materials which compose the samples are relatively brittle.

On the other hand is has become clear that electronic circuitry is indeed affected by deformations and the change of e.g. transistor drain currents is noticable. Drain current variations in the order of a few percent will very likely not disturb the functionality of digital devices, however analogue circuitry might well be affected. A very interesting effect from the basic point of view is that even the direction of the bending axis makes a difference. Perhaps this is related to different variations of the transistor dimensions induced by the deformation.

A next interesting step would be to investigate the performance of devices having different thicknesses. Obviously the tensile stress on the surface of the bent samples gets smaller at the same bending radius, if the samples are thinner. There are some results measured on devices of 28 μm thickness at the Fraunhofer IZM (Munich), which show a much smaller deviation of transistor characteristics. So possibly mechanical deformations are less problematic for devices, which are even thinner than 50 μm.

Moreover, the most important question from the industry point of view is, whether complete ICs will work properly and reliable when facing not only under static, but also dynamic mechanical deformation, like it is the case for smart labels and flexible assemblies. In the final phase of the FLEX-SI project, demonstrators incorporating 50 μm thin IC's will be assembled and tested.

4.3   THIN WAFER HANDLING

Since handling and shipping of ultra-thin wafers is a major challenge in the project, a special work package, handling of ultra-thin silicon - coordinated by Datacon, focuses on proper solutions for these tasks. Expected difficulties that have to be solved are the fragility. Wafers are very sensitive to cracks at the side which is as sharp-edged as a blade, figure 6. Furthermore the bow of the wafers make automatic robot handling difficult. A further difficulty is to buffer thin wafers by means of magazines between two process steps performed on different machines. Since ultra-thin silicon is flexible, the wafer warpage, caused by pure gravity, internal stresses or accelerations during transport, would prevent from using state of art methods for standard thick wafers, i.e. stacking wafers in magazines.

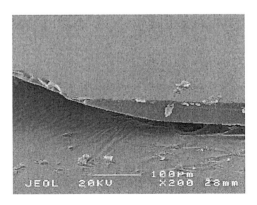

Fig. 6: REM picture of sharp wafer edge.

A major issue in the production of ultra-thin wafers is the process sub chain starting after completion of the rough back grinding process, where the wafer thickness is about 150 μm, and ending with the die bonding process, where ultra-thin chips have to be handled. There are two philosophies concerning ultra-thin silicon handling.

Philosophy 1: Carrierless or flexible carrier based handling

The wafer or chip is either handled without carrier, or a flexible carrier similar to a foil. This kind of handling might have cost advantages, since some process steps might be cheaper or even be omitted.

Philosophy 2: Rigid handling carrier

The silicon is always mounted on a rigid handling carrier, until the die attach or flip chip bonding has been done. The clear advantage of this approach is, that the ultra-thin wafer or chip handling does not differ much from conventional (state of art) handling. Existing equipment can be used. The technical issues arise with mounting and unmounting techniques. Since the rigid carrier may be diced with the wafer, the issue could also be a matter of cost.

4.3.1 Prototype Handling for Ultra-thin Wafers

Within the FLEX-SI project W.S.I. provides the other project partners with processed ultra-thin wafers of 50 μm rest thickness, which are either unbumped or have bumps at 5-20 μm height. The thinning process starts with wafer back grinding down to a thickness of about 150 μm and continues with a special fine grinding and polishing process to reach the target thickness. The bumping is usually done before the back grinding process. W.S.I. uses a handling concept which has been shown to be very reliable for manufacturing ultra-thin wafers with backgrind/polishing techniques. This concept is based on the use of a back grinding tape which is mounted on the wafer's front side, a state of the art method for back grinding technology. The tape serves as a compliant cover for the wafer surface and bumps (if there are any) and protects the wafer surface and bumps from direct contact with the chuck of the back grinding machine, thus preventing non uniform pressure distribution in the vicinity of the bumps. A further function of the tape is to support the ultra-thin wafer with mechanical stability and to reduce mechanical stress. After polishing the wafer down to a thickness of 50 μm, the ultra-thin wafer (mounted on the back grinding tape) is removed by an automatic handler with vacuum gripper from the polishing machine and deposited on a rigid tray.

The handling steps up to this process stage can be done automatically by automated machinery. For ultra-thin wafers in the 50 μm thickness range the next steps of the prototype handling are done manually. The wafer is picked manually with tweezers from the tray and placed on a vacuum chuck with the reverse side down. Standard procedures are used for removing the back grinding tape, i.e. mounting a piece of detaping tape on it and peeling the back grinding tape off the wafer. After releasing the vacuum, the wafer is removed from the chuck with the tweezers and placed in a shipping box.

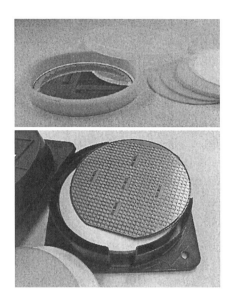

Fig. 7:
Shipping boxes for
ultra-thin wafers

One kind of shipping box is a special cylindrical box with foam padding around the edge and at the bottom, figure 7. Inside the box the ultra-thin wafers can be stacked, separated by paper discs. The top of the wafer stack is covered with an additional foam pad, which presses the paper separated wafers close to each other, when the box is closed. With this kind of shipping package, ultra-thin wafers can be shipped very reliably, although it is not trivial to unstack the wafers by automated robot handling (wafers have been known to stick together from case to case).

4.3.2 Carrierless or Flexible Carrier Based Handling

The FLEX-SI consortium focuses on handling concepts, where the silicon wafer or chip have neither a carrier support nor is supported by a flexible carrier (e.g. foil). Prototype investigations have shown, that e.g. ultra-thin wafers without any carrier support can be reliably shipped, can be mounted on film frame carriers for dicing with standard dicing equipment, and separated dies can be ejected and handled by die bonders (flipped or non flipped) without the need of a carrier support, figure 8. The wafers, processed at the Fraunhofer-ISIT fab, thinned by W.S.I. have been mounted and diced at ISIT. Die bonding experiments have been investigated both at ISIT and Datacon.

Fig. 8: Die bonder, picking 50 μm thick chip from dicing foil (chip without carrier support).

4.3.2.1 Concept A – Cluster Tool Handling

The first concept overcomes the need for a magazine buffering between wafer thinning and mounting on a dicing tape. A key principle of this approach is, that the ultra-thin wafer is always sucked by means of vacuum either on a handler or chuck in order to keep the wafer flat and preventing from warpage. A further key principle is, that the thinning machine and the wafer mounter are located close to each other, so that the ultra-thin wafer can be passed directly from the robot handler of the thinning machine, which touches the wafer on the passive side, to the robot handler of the wafer mounter, touching at the active side. This is a typical approach with cluster tools where a single tool combines the capabilities of wafer thinning, optionally detaping, and mounting the wafer on a film frame carrier.

Thereby the ultra-thin wafer is assumed to be supported by a protective tape on the active side. This provides the wafer from being damaged by the wafer mounter's robot handler

4.3.2.2 Concept B – Sandwich Buffering

A disadvantage of the *cluster tool handling* concept is, that there is no buffering between thinning machine and wafer mounter, which means, that both machines have to work in an absolute synchronous mode. An alternative approach is called *sandwich buffering*, which bases on the method of putting the ultra-thin wafer between two rigid discs. These discs can be made of silicon. Such a constructed sandwich can be stored in a magazine.

4.3.2.3 Concept C - Film Frame Carrier Based Buffering

The basic idea of the sandwich concept is to prevent the ultra-thin wafer from warping, when it is stored in a magazine. An alternative approach is to prepare a magazine with film frame carriers, where a sticky handling tape has been mounted in a previous step. After wafer thinning the handling robot takes the wafer out from the thinning machine touching the ultra-thin wafer on the passive side and puts it with the active side, which is covered with the protective tape, down to the sticky foil on the film frame carrier.

4.3.2.4 Concept D – Thinning with Film Frame Carrier

With concept C the mounting of the ultra-thin wafer on the film frame carrier supported handling tape has to be done by the output handler of the thinning machine. Although this step does not seem to be very crucial, there is no indication up to now, whether troubles could occur or not. An alternative approach is to do the mounting before the thinning process, i.e. when the wafer is in the thin state (about 150 μm). This implies that standard wafer mounting equipment can be used. As a big trade off the thinning machine must be capable of processing wafers mounted on film frame carriers. As with concept C a final cross mounting step is necessary.

4.3.3 Rigid Carrier Based Handling

4.3.3.1 Basic Approach

In contrast to the wafer handling concepts introduced in the previous sections a second type of approach is based on rigid handling carrier support. A rigid carrier support, where the carrier dimensions do not exceed the wafer dimensions, has the big advantage that the supported wafer mechanically behaves like a conventional thick wafer. This implies that all existing equipment and handling techniques can be used for wafer processing.

The mounting can be done either by double sided sticky tape or depositing an adhesive layer either on the wafer or on the carrier by means of spinning methods. Since the carrier is mounted on the active side for back grinding and thinning, a cross mounting on film frame carrier has to be done for wafer presentation to the dicing machine.

4.3.3.2 The AWACS Process Flow

The AWACS process flow (Advanced Wafer And Chip Supply) is an idea from Philips Semiconductors (patent filed) and proposes a flow in which the wafer is always supported by a rigid substrate. The advantages are:

- no equipment adaptions are necessary
- the method can be easily integrated in existing production flows
- testing after thinning is possible
- the investment costs are low since existing equipment with standard handling can be used
- appropriate for all wafer diameters (also for 300 mm wafers)
- independent of the thinning technology (grinding, polishing, spin-etching, plasma etching)

The AWACS flow comprises the following steps:

1) Initial product is a thick wafer (maybe bumped)

2) A double sided adhesive foil is laminated on the wafer front side

3) A substrate A (handling carrier) is mounted on the foil

4) Wafer back side is ground and etched

The intermediate product is a stack consisting of the ultra-thin wafer, which is mounted on UV transparent substrate A by means of a double sided UV foil. Since the front side of the wafer must be accessible for testing, a cross mounting step to a different handling carrier requires to be performed. By philosophy of the AWACS flow this is also a rigid carrier (substrate B).

5) Double sided adhesive foil laminated to the back side

6) Application of substrate B

7) UV release of foil between substrate A and wafer front side.

8) Delamination of UV foil and substrate A

4.3.4 The Role of Tapes and Adhesives

As described in the section above there are many steps of releasing a foil or a rigid carrier from the wafer. The cross mounting step in the rigid carrier based handling seems to be especially crucial, since there are no standard procedures to release the rigid handling carrier from the wafer. To make life easier, an adhesive type could be used, where the adhesive force could either be lowered by UV light or by temperature treatment. In case of a UV type adhesive one should also keep in mind, that the handling carrier must be transparent for UV radiations, as it is the case with e.g. a quartz carrier. Japanese companies have been known to offer special foil types with UV or thermal release capabilities.

4.3.5 Recommendations and Conclusions

Ultra-thin wafers and dices are handled on film frame carriers using special tapes with low residual adhesion (UV or thermal release tape). Wafer feeder should be used to feed ultra-thin dice to die bonder resp. flip chip bonder to minimize die handling steps. Waffle packs should be avoided also because of particle problems. Si splinters from the die backside might cause die cracks when they are located in the bondline.

For the automatic assembly of ultra-thin chips optimized die bond equipment is required with adapted tooling and special release tapes.

A major issue in processing ultra-thin wafers is the handling concept, for which there were special research activities within the FLEX-SI consortium. There is a prototype handling scenario which makes it possible to produce and ship ultra-thin wafers. There are two major approaches for ultra-thin wafer handling: 1) Carrierless or flexible carrier based handling and 2) rigid carrier based handling. For each approach there are different methods possible. The feasibility of these methods have to be investigated in future work.

4.4   ULTRA-THIN CHIP STACKING

Apart from low-profile flexible flip chip assemblies, ultra-thin chip stacking in chip-on-chip technology for system-in-package is another industrial application of the FLEX-SI project. For the chip-on-chip assembly with ultra-thin dice the die bonding is the crucial process step. Therefore, the die bonding of ultra-thin chips an issue to investigate:

- different thinning and stress release techniques affect die strength
- no standard handling of thinned wafer, storage, shipping
- dicing tape affects die pick process yield
- direct impact of wafer dicing quality on chip cracking
- chips bend under mechanical load (pick, place)
- adhesive application difficult, contamination of pick up tool
- compliant adhesives for flexo elastic assemblies necessary
- optimization of bonding equipment necessary

4.4.1 Application of Adhesive for Die Attach

In principle, the die bond adhesive can be applied on the substrate as a paste
or as a preform, or it can be applied on the chip backside, also as paste or
preform, see figure 9.

Application of die bond adhesive
      on the substrate: preform   on chip backside:
           • dispensing   • dip transfer
           • stamping   • transfer tape
           • stencil printing

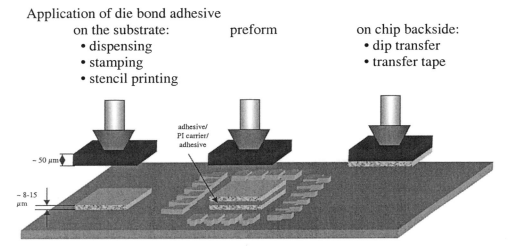

Fig. 9: Techniques for adhesive application for die attach.

In the FLEX-SI project the focus was put on the techniques where the
adhesive for die attach is applied on the backside of the chip. In the
following, two techniques for the application of adhesive on the chip will be
presented in detail including a benchmarking of the performances:

- Dip transfer of adhesive
- Transfer tape on the chip backside

4.4.1.1 Dip transfer of adhesive

In the dip transfer process, patent pending by KRT technologies, the chip is
dipped in an adhesive film through a mesh in a special dipping station in the
die bonder, see figure 10. The mesh can have a structure to reduce the
adhesive coverage or fillet. The tool size has to be adapted to the chip size.
The mesh has the function to support the release of the chip after dipping it
into the adhesive film.

Figure 11 shows chip-on-chip assemblies with test chips where the die bond
adhesive has been applied by the described dip transfer process, the
advantages and disadvantage of this method to apply die bond adhesive are
presented in table 3.

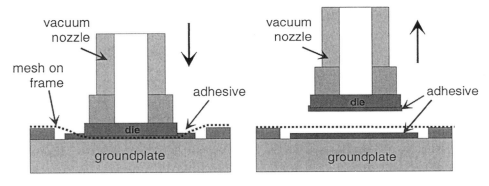

Fig. 10: Principle of the adhesive dip transfer through a mesh

Fig. 11: Chip-on-chip test assemblies with adhesive dip transfer

Pros	Cons
+ full area adhesive wetting + no height control required + broad chip geometry + defined fillet and bond line thickness + excellent for chip stacking	− automatic process difficult due to adhesive build up and cleaning of mesh − voiding possible with large dice − chip crack possible − special tooling required

Table 3: Advantages and disadvantages of the dip transfer process

4.4.1.2 Transfer tape on the chip backside

In this technique the die bond adhesive is applied as a film on wafer-level. After dicing of the wafer each chip is carrying an adhesive label on the backside. Figure 12 is showing the principle of the transfer tape use, as well as an ultra-thin chip stacking assembly with 50 μm chips realized with adhesive transfer tape. The advantages and disadvantage of the transfer tape method to apply die bond adhesive are presented in table 4.

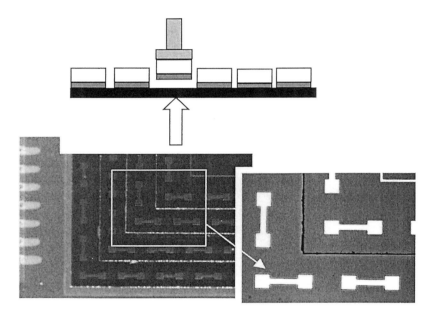

Fig. 12: Use of adhesive transfer tape for ultra-thin chip stacking

Pros	Cons
+ high speed process – no adhesive has to be applied on the machine	– small adhesive spectrum available
+ defined adhesive thickness	– no electrical conductivity
+ excellent for chip stacking	– substrate heating necessary
+ no tilt, no voiding	
+ no contamination of the tool	

Table 4: Advantages and disadvantages of the transfer tape process

4.4.1.3 Benchmarking of adhesive transfer tape vs. dipped epoxy

In a comparative evaluation of the dip transfer process using epoxy adhesive and the transfer tape process, the following tests were performed:

- bond strength by die shear test
- voiding analysis by non-destructive test methods

Bond strength analysis: die shear test

The die shear tests have been performed according to the MIL-STD883, method 2019.5 using chips of full thickness. Both die bond types passed the MIL standard specifications. Figure 13 shows the results of the die shear tests on different surfaces (epoxy, solder resist, flash Au) for the dip transfer of epoxy (left) and the adhesive transfer tape (right). The bond strength of the epoxy die bond (dip transfer process) is higher especially on solder resist and flash-Au surfaces. But the bond strength of the transfer tape should be sufficient for the ultra-thin chip stacking.

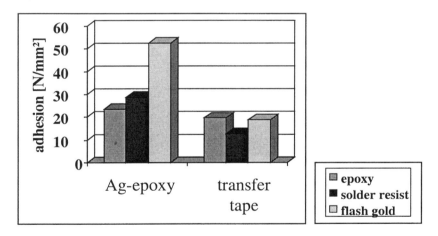

Fig. 13: Die shear test results of the two adhesive application techniques on different surfaces.

Voiding analysis by scanning acoustic microscopy and x-ray transmission

Figure 14 shows the results of the non-destructive analysis of voids in the adhesive layer for die attach by scanning acoustic microscopy (left) and x-ray microfocus transmission (right). Both techniques show very good correlation in the case of the dipped (filled) epoxy adhesive revealing some larger voids. The transfer tape causes no x-ray signal, the SAM picture shows no voids.

Scanning Acoustic Microscopy (SAM):   X-ray Transmission:

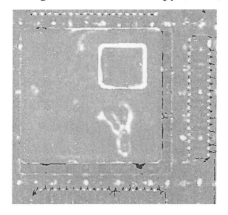
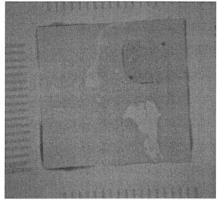

Fig. 14:
Comparison of voiding using
SAM and x-ray analysis
  top: dipped epoxy
  buttom: transfer tape
The transfer tape causes no
x-ray signal.

4.4.2 Die Pick-up Alternatives

The process of picking the die from the tape of the film frame carrier is the most crucial step in automatic die bonding of ultra-thin chips because of the risk of die fracture. Several principle approaches with penetrating or non-penetrating neddles as well as needleless concepts are possible. Figure 15 presents the basic principles of the different concepts.

Apart from the ejection needle system, the choice of the tape is an key issue for high yield die bonding. UV tapes are more or less standard for working with thin wafers. Special tapes with thermal release at a defined temperature are available e.g. the Revalpha tape from Nitto.

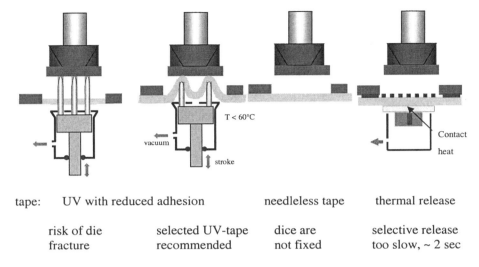

tape: UV with reduced adhesion   needleless tape   thermal release

        risk of die   selected UV-tape   dice are   selective release
        fracture   recommended   not fixed   too slow, ~ 2 sec

Fig. 15: Alternative solutions for die ejection from the adhesive tape of the film frame carrier

The following demands and equipment features for automate die bond equipment working with ultra-thin chips have been realized by Datacon within the FLEX-SI project:

- flexible machine platform with automatic wafer handling
- combined flip chip and die attach capability
- various adhesive application techniques
- soft and synchronous die ejecting
- automatic tool exchange
- heated underboard support and tools
- synchronous ejecting: pick-up tool and flip tool are moving at synchronized speed during ejecting
- soft take over of die
- needle systems and tools adapted to chip geometry
- slow motion programmable
- defined bond force
- z - positional bonding and dispensing

The prototype flip chip die bonder for ultra-thin silicon which has been developed within the FLEX-SI project is evaluated under close to production conditions in a cooperation between Datacon and ISIT with two user

companies of ultra-thin silicon from smart label and power electronics applications.
Figure 16 shows an ultra-thin chip stacking assembly with 50 μm test chips. The die bond has been realized with adhesive transfer tape.

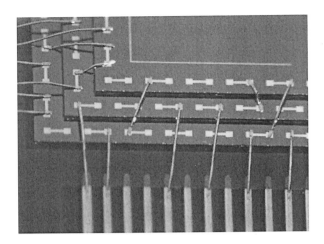

Fig. 16: Chip-on-chip test assembly with ultra-thin dice,
die attach realized with adhesive transfer tape

4.4.3 Conclusions

Automatic die bond and flip chip assembly of ultra-thin chips with 50 μm thickness was successfully performed on a modified Datacon apm 2200 flip chip die bonder on following conditions:

- use of tapes with reduced release adhesion
- chip sizes: 1.5 x 1.5 mm^2 up to 15.0 x 15.0 mm^2
- feeding 6 inch wafer with 50 μm thickness on frame in wafer magazine
- non-penetrating needles, needle system and tools tuned one to another
- soft needle speed; tool and needle movements synchronized

4.5 LOW-PROFIL ASSEMBLIES USING ULTRA-THIN ICs

4.5.1 Flip Chip On Flex

The small-gap flip chip assembly of ultra-thin dice with low-profile bumps on thin and flexible substrates is enabling thin and flexible multi chip modules. To achieve a reliable flexible module the chips should be placed in the neutral layer of the bending.

Figure 17 shows an example of flip chip on flex assembly of ultra-thin dice for a smart label module.

Fig. 17: Smart label modules (flip chip on flex)

4.5.1.1 Process flow for flip chip assembly with ultra-thin ICs

For the flip chip assembly with ultra-thin chips the FLEX-SI project followed the approach of wafer bumping <u>before</u> thinning, figure 18. The main reason for this decision is that there is no industrial process for the bumping of ultra-thin wafers. As a consequence, the wafer thinning process has to be performed with bumped wafers.

The following challenges in the flip chip assembly of ultra-thin chips especially for solder flip chip have been met:

- no flux dip possible due to flat bumps
- capillary flow underfill after reflow difficult
- tool contamination due to underfill squeeze out likely
- die can float after placement or during reflow
- commercial available solder masks too thick for pad definition
- warped dice dependent on IC technology

92

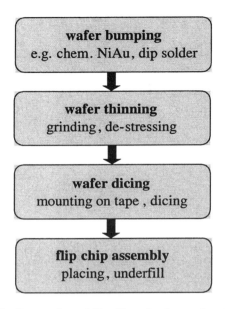

Fig. 18: Process flow of flip chip technology using ultra-thin ICs

4.5.1.2 Under-bump-metallization (UBM) and bumping

In the FLEX-SI project only low-profile bumps were considered for flip chip assembly with 50 μm thin chips, on the one side because of the limitations from the thinning process for bumped wafers, on the other side because of the overall height of the flip chip assemblies. Stencil printed solder bumps with a typical bump height of 100 μm and Au stud bumps with a typical height of 40 – 80 μm are not useful in this application. Therefore, the following bumping techniques are applied:

- electroplated Au bumps (typ. height 15 – 20 μm)
- electroless NiAu bumps (typ. height 5 – 20 μm)
- electroless NiAu + solder bumps (typ. height 20 – 25 μm)

The electroless deposited NiAu under-bump-metallization can be directly used as low-profile bump e.g. for a flip chip process with electrical conductive adhesive joining.

4.5.1.3 Solder bumping

The solder flip chip process requires pre-soldered bumps with limited height resp. solder volume. For this application a dip solder transfer technique is used to apply different solder alloys in an appropriate solder volume on the electroless NiAu UBM. Eutectic SnPb-solder, low-temperature Bi50Pb31.3Sn18.7 solder as well as Pb-free SnAg3.5 has been evaluated.

Wafer fracture did not occur during the dip solder transfer process. Figure 19 shows a test chip with NiAu/solder bump after the dip solder transfer. No additional reflow is necessary after the dip coating to form rounded solder bumps. Good results have been achieved with the dip solder transfer process operating in nitrogen inert gas atmosphere or with a solder bath overfilled with an organic medium which improves the heat transfer to the wafer and provides an oxygene-free environment. The dip time in the solder is about 2 sec.

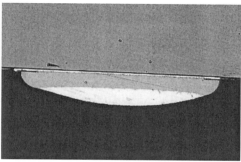

Fig. 19: Low-profile NiAu/solder bump after dip solder transfer

4.5.1.4 Substrates for low-profile flip chip assembly

Apart from flexible polyimide substrates printed circuit boards (FR4) with precision resist, 25 μm Cu, 4 μm Ni / flash-Au were used in two realizations, figure 20:

- solder mask defined contact pads
- non-solder mask defined contact pads

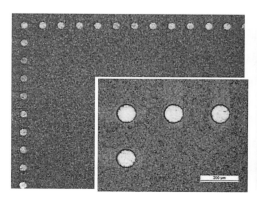
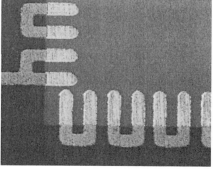

Fig. 20: PCB substrates for flip chip with ultra-thin dice.
left: solder mask defined pad right: non-solder mask defined pad

Although the solder resist film is very thin (20 μm) it causes problems in the solder mask defined design where the solder mask is also located under the chip. If low-profile bumps are used, the chip will come in contact with the resist, whereas the solder caps of the bumps do not contact the pads on the substrate. Furthermore, the small gap between chip and substrate is further reduced, hindering the flow of underfill. Therefore, if solder mask defined pads are used, bump height and resist thickness have to be adjusted, and pre-applied no-flow fluxing underfill is preferred to standard capillary flow underfill material.

4.5.1.5 Flip chip attach

Flip chip test assemblies with bumped test chips have been produced using the solder flip chip process. Test chips of 50 μm thickness and 4.75 x 4.75 mm² size with NiAu/solder bumps applied by the described solder dip transfer process were used. Figure 21 shows an ultra-thin flip chip on a FR4 printed circuit board. The solder flip chip process has been studied with different underfill techniques:

- solder flip chip with post applied underfill (unfilled type)
- solder flip chip with post applied underfill (filled type)
- solder flip chip with pre-applied no-flow fluxing underfill

Fig. 21: *Ultra-thin solder flip chip with 50 μm test chip (0402 SMD for height comparison)*

The underfilling of ultra-thin solder flip chips cannot be performed like with standard flip chip, as the resulting gap may be below 20 μm. The work is directed to pre-applied underfill with a protection of the die placement vacuum tool against polymer contamination.

Figure 22 shows a cross section of the thin flip chip on board assembly using the low-profile NiAu/solder bumps. The insert shows a single contact in detail.

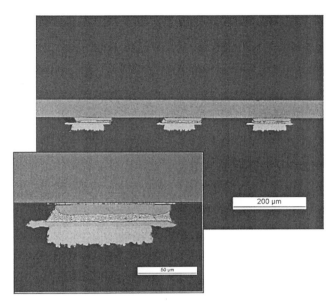

Fig. 22: Cross section of an ultra-thin solder flip chip

4.5.1.6 Summary and Conclusions

Low-profile flip chip assemblies have been realized with ultra-thin Si chips using a solder flip chip technology. As low-profile bumps electroless NiAu/solder bumps were used where the flat solder cap is applied by a dip solder transfer process. Attention has to be paid to the substrates as standard solder mask designs do not work. Pre-applied no-flow fluxing underfill is preferable compared to standard capillary flow underfill because of the small underfilling gap.

4.5.2 Chip-In-Board Technology: Flexible IMB Module

In Helsinki University of Technology a solderless interconnection and packaging technique has been developed. In this Integrated Module Board (IMB) technique active as well as passive components are embedded inside an organic substrate. The active components are embedded using a Chip-in-Board (CIB) technique and the electrical contacts are realized using a fully additive copper deposition process. The integrated passive components are manufactured simultaneously with the HDI build-up wiring.

The principal IMB steps are schematically presented in figure 23. In the first step the holes for the active components are drilled or stamped depending on the substrate used (Fig. 23(a)). The first signal layer can be patterned to the core board using, e.g. print & etch process before embedding the chip. In the second step an adhesive film is laminated over the substrate and the bare chips are placed accurately on the film (Fig. 23(b)). Then the chips are encapsulated using a molding polymer (Fig. 23(c)) that is cured after which the adhesive film is removed. Subsequently the surface is catalyzed for the chemical copper deposition (Fig. 23(d)). A photosensitive dielectric material is coated over the surface and the openings for chip I/Os as well as the first layer signal traces are formed by the photolitographic process. The electroless copper is then deposited to the openings to form concurrently electrical contacts and the first signal layer (Fig. 23(e)). This approach enables the fabrication of all layer via structures.

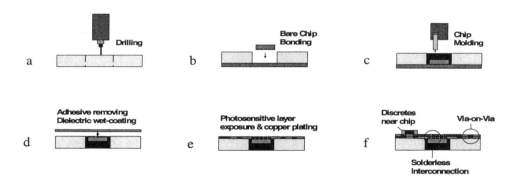

Fig. 23: Major IMB process steps

IMB technique is also capable when embedding thin active components in flexible substrates. In these flexible modules also the molding epoxy used is elastic resulting to completely flexible structures.

In the FLEX-SI project a 0.1 mm thin FR4 was used as substrate material. The thickness of embedded flexible ICs was 50 μm. These ICs contained a daisy chain structure with 72 I/Os in 250 μm pitch. These I/Os were coated with a 5 μm Ni/Au bumps. A polyimide adhesive tape was attached over the drilled hole and a mask was placed below the substrate. A flip chip bonder was used to align the active components to the contact pads of the mask and to place them onto the tape. The mask was removed and the active components were embedded with a molding epoxy. After epoxy was cured the polyimide tape was removed. The active surfaces of the chips with the I/Os were now at the same level with the surface of the substrate, as figure 24 shows.

Fig. 24:
Ultra-thin test chip embedded
in a flexible FR4 substrate.

Interconnections and wiring were fabricated with a fully additive multilayer printed wiring board process similar to that in figure 23. In the first dielectric layer only the microvias for I/Os were formed. Figure 25 shows the 2nd dielectric layer with the photolitografically fabricated openings for daisy chain 2nd level interconnections. The completed flexible IMB module is presented in figure 26.

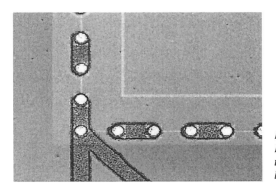

Fig. 25:
Flexible module after
the 2nd layer photo-
litographic process.

Summary

The IMB-technique being developed at Helsinki University of Technology enables very high density integration of active and passive components into organic substrates. Thinned active components (50 μm) were integrated into the flexible substrate using elastic molding polymer. Electrical interconnections were fabricated with the fully additive solderless process, where liquid photoimagible polymer was used as dielectric layer and high density wirings and interconnections were produced using electroless copper deposition. Since the module consists of flexible elements the resulting structure provides enhanced operational reliability.

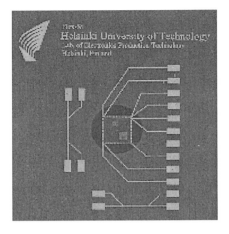

Fig. 26: Flexible IMB module after electroless copper plating and interconnections formation.

4.6 SUMMARY AND CONCLUSIONS

The results of the FLEX-SI project can be summarized as follows:
- wafer thinning to 50 μm with de-stressing works well
- tested devices are functional on thinned and bent wafers
- drain current changes in ultra-thin and static bent state
- bump wafers before final thinning
- use low-profile bumps (e.g. electroless NiAu, dip solder, galv. Au)
- use film frame carrier with special tapes to handle ultra-thin wafers and chips
- standard wafer dicing is possible with special tape and blade
- optimize pick & place for controlled touch down and force
- transfer tape type die attach is preferable especially for ultra-thin chip stacking
- match pick tool and needles (non-penetrating needles, needleless)
- pre-applied (no-flow) underfill is preferable
- standard solder mask designs do not work
- IMB technique works with ultra-thin chips

In the FLEX-SI project existing industrial processes e.g. for wafer handling, dicing and die bonding were used as a starting point for ultra-thin wafer processing instead of creating totally new processes which are not compatible to the existing infrastructure in industry.

The FLEX-SI approach of modifying and adapting existing processes to work with ultra-thin silicon made it possible to provide a full chain of industrial processes for advanced packaging with ultra-thin chips, from the wafer thinning and dicing to functional demonstrators e.g. in flip chip or chip-on-chip technology.

Ultra-thin packaging solutions and flexible electronic assemblies will enable new products in the field of information and mobile telecommunication technology. Examples are wrist-watch communicators, a medical control unit which is carried at the body, concealed electronics as a protection against larceny, an integrated sensor module for monitoring the load affecting humans at their place of work, flexible identification systems attached to goods to enable the tracking of goods flow including waste disposal control and improving anti-counterfeit techniques.

Acknowledgement

The work has been carried out under the IST-99-10205 Project FLEX-SI ´Ultra-thin packaging solutions using thin silicon´ funded by the European Commission. The authors gratefully acknowledge their support.

Furthermore, we would like to thank the co-authors from the FLEX-SI consortium, especially Christian Brugger (Philips Semiconductors), Hugo Pristauz, Christoph Scheiring and Gerhard Hillmann (Datacon Semiconductor Equipment) as well as Vesa Vuorinen and Jorma Kivilahti (Helsinki University of Technology).

References

T. Harder, W. Reinert
´Low-profile flip chip assembly using ultra-thin ICs´, 13th European Microelectronics & Packaging Conference (IMAPS), 30.05. - 01.06.2001, Strassbourg

H. Pristauz
´Handling concepts for ultra-thin wafers´, 13th European Microelectronics & Packaging Conference (IMAPS), 30.05. - 01.06.2001, Strassbourg

G. Hillmann, W. Reinert (ISIT)
´Automatic assembly of ultra-thin chips´, Thin Semiconductor Devices – Manufacturing and Applications, Fraunhofer IZM, Munich, December 2001

T. Harder
´Low-profile and flexible electronic assemblies using ultra-thin silicon – the European FLEX-SI Project" at the planary session on foldable flex and thinned silicon at the International Conference on Advanced Packaging and Systems (ICAPS 2002, March 2002 in Reno)

T.F. Waris, J.K. Kivilahti
´Manufacturing of flexible integrated module boards´, IMAPS Nordic 2002

Chapter 5

THIN CHIPS FOR FLEXIBLE AND 3D-INTEGRATED ELECTRONIC SYSTEMS
Thinning, Dicing, Handling and Assembly Processes for Thin Chips built an Integrated Approach

Karlheinz Bock, Michael Feil, Christof Landesberger
Fraunhofer Institut for Reliability and Microintegration – Munich Division (IZM-M)

Abstract: Thin chip work at the IZM aims to develop an integrated approach from the Thinning of product wafers over related chip separation and handling techniques to the application of useful assembling processes to bring the thin chip safely into the product and to allow a reliable operation. After an introduction to thin chip applications the chapter focuses on the development of thinning processes and explains the different thinning techniques applied. In a strong correlation the handling and wafer preparation is described in detail. In a further part the correlated assembly processes are presented in a relation to possible product or application scenarios. Thinning, Dicing, Handling and Assembly Processes for Thin Chips are developed at the Fraunhofer IZM-M together in an Optimized and Integrated Approach. The choice of a thinning process in many cases defines already the boundary conditions of the separation handling or assembly process of a product. Therefore having the product in view none of the processes can be developed in an isolated manner. The chapter finishes with a short outlook and the introduction of reel to reel processing of systems with thin chips and manufacturing of low-cost thin chip products in the IZM reel to reel application center.

Key words: Thinning, Grinding, spin etching, wet chemical etching, CMP, chemo-mechanical polishing, cleaning, lamination, carrier, adhesives, ICA, NCA, ACA, flip-chip assembly, iso-planar interconnect, 3D-integration, chip on flexible substrates, reel-to-reel processing, polymer electronics, polytronics

1.   APPLICATION POTENTIAL OF ULTRA THIN ICS

Integration of new functions into high performance portable equipment is one driving force behind the increase of component and interconnection densities and thereby the decrease of size of electrical assemblies and their components like chip and substrate thickness as well as interconnection density. This adds to the importance of mechanical compatibility between components and high density boards, since stresses of either thermal, electrical or mechanical origin can cause serious reliability problems. Thin chips in rigid orflexible assemblies can help to increase the component packaging density, to decrease the systems volume or to reduce the topology and to increase the reliability as shown for thin flip-chip assemblies. When wafer thinning is extended to ultra thin chips with a remaining thickness of 10 - 30 μm, a large field of new technological possibilities and completely new applications appear. In this thickness range silicon becomes pliable and tolerant regarding mechanical stresses.

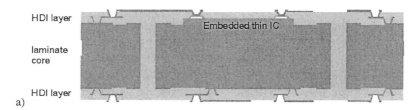

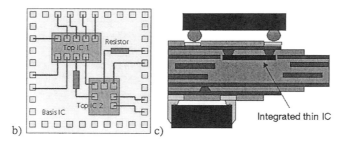

Figure -1. Packaging concepts using ultra thin ICs a) Chip in Polymer, b) Chip on Chip, c) Chip in multiplayer PCB

The reduction in IC thickness leads to very compact packaging concepts. Diverse 3-D and high density assembly techniques can be simplified. Some examples like system-on-chip (SOC) or chip-on-chip (CoC) solutions, chip in polymer or paper (CiP) are given in fig 1 [1].

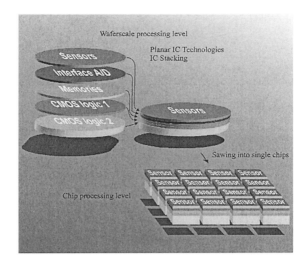

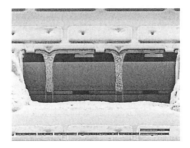

Figure -2. Vertical system integration on wafer level
a) schematic sketch, b) cross section of a stacked test sample using Focused Ion Beam (FIB)

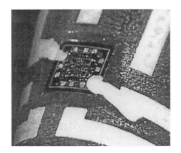

Figure -3. Strain gauge sensor

Ultra thin silicon allows vertical integration on wafer level, too. By this assembly technique stacking of logic, memories, sensors and the sensor-logic interface will be possible(s fig. 2) or wafers with different or simplified semiconductor processes (CMOS, bipolar, etc.) can be integrated on a single silicon wafer [2, 3].

Wafer thinning delivers a new approach for MEMS fabrication. Fig. 3 shows a thinned chip with implanted piezoresistors acting as strain gauge. These chips can be mounted on curved surfaces. Commonly anisotropic etching methods are used to fabricate membranes. With ultra thin silicon the whole element has the same thickness. Therefore, the whole chip is flexible and bent according to the working pressure or force. It can be mounted on a separate membrane body, which is customized in shape and material according to the requirements of the mounting place and the surrounding media, which it comes in contact with(s. fig. 4) [4]. Advantages are simplified processing, cost reduction, flexible, clinging on to any surface, very low weight and mass.

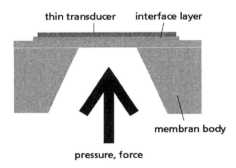

Figure -4. Scheme of a force sensor with separated membrane body and transducer

Another application example are optical reproduction systems. Modern systems use electronic photo line or area detectors. They can built up only in one plane. Therefore, a sharp reproduction requires extensive steps for the correction of reproduction errors of lenses or telescopes. A reflecting telescope as used in space application has always a spherical focus. The correction expenditure could be much simpler using a ultra thin detector, bended according the focal radius as sketched in fig. 5 [5].

A very wide field for application of ultra thin chips are low cost products like smart cards and all kinds of RFID labels (s. fig 6) [6-8]. A extremely tin chip, embedded in paper or polymer sheets is not palpable and there difficult to detect. Passive transponders may be used for security applications as "electronic watermark" or in the logistic of goods distribution and book inventory. Embedded in tickets, they may simplify the access control and make more difficult forgery. As a electronic business card, they will give your partners or customers not only the standard information of a normal one but additional information about your company, products etc. or it will contain a joke.

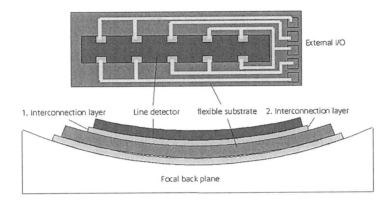

1. Interconnection layer   Line detector   flexible substrate   2. Interconnection layer

Figure -5. Scheme of a optical detector, bended according the focal radius

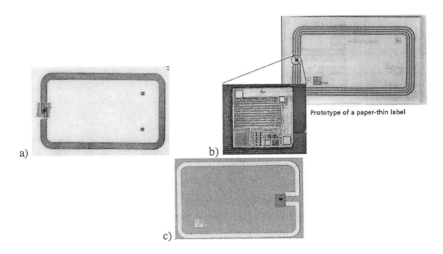

Figure -6. Examples of RFID label
a) RFId-label with etched copper coil, b) RFId-label with a coil consisting of a screen printed polymer paste on paper,
c) RFId-label containing only one turn for short range application

Active transponder with her own energy supply and containing a timer and sensors may be used as data loggers. In the case of a temperature sensor, it monitors the course of temperature over the time.

2. THINNING AND DICING OF SEMICONDUCTOR SUBSTRATES

2.1 Wafer thinning

Most often used wafer thinning technique is backside grinding. To avoid damage of integrated circuits a protective tape is laminated to the front side of a device wafer. Backside grinding is a fast and precise thinning technique. Removal rates up to 300 μm per minute are used for production purposes. As the amount of silicon removal is measured in-situ by thickness gauges the target thickness can be controlled by a few micrometers. Ground wafers exhibit a total thickness variation (TTV value) in the range of 1 to 3 μm for 150 mm wafers. During grinding wafer backside is abraded mechanically by a rotating wheel carrying embedded diamonds. Wafers are cooled by flushing de-ionized water. A typical grinding process consists of two steps: rough grinding and subsequent fine grinding. Due to high pressure which may occur at the contact area of diamond crystals, the backside of silicon wafer is damaged to a certain extent. For thinner wafers backside stress leads to highly bent silicon substrates and increases the risk of wafer breakage. This problem is overcome by adding a stress relief process after grinding. Suggested techniques are wet-chemical etching [9], CMP polishing [10] or plasma dry etching [11, 12]. An overview on current thinning and stress-relief processes are given in table 1.

An isotropic etchant HNO_3, HF, H_3PO_4 is often used for wet-chemical removal of silicon. Today so called "spin-etch" process has become a standard technique for wafer backside treatment. Removal rate depends on HF concentration and temperature of etchant; typical values are in the range of 10 to 40 μm per minute. Damaged backside layers of ground silicon wafers can be removed effectively. Process ends up with a mirror like backside.

Plasma etching of silicon using fluorine radicals is a well known process since decades of years. However, isotropic backside etching of silicon wafers for thinning purposes is a recent development, which was driven by a few companies within the last years. Dry etching offers interesting advantages for wafer thinning. It does not induce mechanical stress to the wafer and it allows etching of micro structured wafers like MEMS devices. Dry etching of silicon is an exothermal process. This means high etching rates cause high wafer temperatures. Therefore the application of a plasma process to taped wafers renders more difficult. This problem could be solved in the future by carefully adapted etching processes, enhanced chuck cooling or new temperature stable techniques for wafer support.

Table -1. Overview on current wafer thinning techniques and relevant process parameters.

	Grinding	Spin - etching	Dry – etching	Polishing
Type of process	Mechanical abrasion	Wet-chemical etching	Plasma, Reactive ions	Chemical and mechanical removal
Process medium	Diamonds in ceramic wheel	HF + HNO$_3$ + additives	SF$_6$, NF$_3$, XeF$_2$	Slurry: SiO$_2$ grains in soft etchant
Removal rate	300 μm/min	10 ... 40 μm/min	3 ... 30 μm/min	< 3 μm/min
TTV: total thick-ness variation	0,5 ... 3 μm/min	5 .. 10 % of removal	n. a.	< 1 μm
Process temperature	cool	30 ... 40 °C	50 ... 300 °C	30 ... 40 °C
application	Thinning	Stress-relief	Stress-relief, MEMS thinning	Surface finish, planarization

As further possibility for a stress relief process after grinding chemical mechanical polishing (CMP) was successfully applied to thin wafer fabrication. Also integrated process concepts for wafer grinding and polishing are offered by equipment suppliers.

It can be stated as a final remark, that various thinning techniques are available to reduce wafer thickness down to a few micrometers. However, due to increasing wafer bow and a dramatic increase of risk of wafer breakage, things get serious when wafer thickness below 100 μm should be reached. Furthermore for thin wafers also the dicing technique has major influence on the material strength of a thin die.

An overview on existing handling techniques for thin wafers is given in the next section.

2.2   Handling techniques for thin wafers

Free standing wafers can be handled by standard equipment (robot systems, end effectors) for a substrate thickness down to approximately 100 μm for a 150 mm wafer and down to approx. 200 μm for a 200 mm wafer. This is also applicable to wafers coated by a protective front side tape. For thinner wafers the risk of breakage is strongly increasing. Thin wafers may be bent by several millimeters due to specific front side processing (e. g. stress in layers). Therefore handling of thin wafers into or out of a standard cassette is getting more and more difficult (see fig. 1). Any robot system has to be carefully adjusted to the wafer shape and requires additional sensor functions for detection of wafers in the cassette as well as at the end effector. Special tool modifications are offered by various equipment companies. Using such dedicated equipment handling of free standing thin wafers may tolerate wafer thickness down to approx. 80 μm.

Figure -1. Example of a 150 μm thin silicon substrate, diameter 200 mm, placed in a standard wafer cassette. Large values of wafer bow prevent safe robot handling of thinned wafers

Within the wafer thickness range of 50...80 μm the possibility of handling wafers is depending on following boundary conditions: type of product (specific fabrication steps), degree of automation, possibility for manual support, special tool adaptations and of course wafer diameter.

Most problems in handling thin wafers can be overcame when carrier substrates are introduced to support thin wafers during processing and handling. In this case device wafers are reversibly stacked to a rigid carrier substrate. Today this process scheme is widely used in fabrication of brittle III-V compound semiconductors (e. g. GaAs, InP). Temporary adhesion of device and carrier wafer (e.g. glass or sapphire) is often provided by thermoplastic adhesives or wax. Further possibilities for reversible mounting of carrier substrates are application of double side adhesive tapes, which reduce or loose their adhesion after UV irradiation or thermal treatment.

Handling thin wafers by carrier substrates offer the following advantages: risk of wafer breakage is practically eliminated and fabrication can be easily extended to lower wafer thickness as well as to larger wafer diameters. Costs for additional process step "carrier bonding" have to be compared with the required amount of financial invest if special equipment modifications have to be adapted at each site of fabrication flow.

It can be concluded that the decision between handling of free standing thin wafers and handling by carrier substrates is finally ruled by the value of product wafers and the risk of wafer breakage.

Table -2. Possible handling strategies for thinned silicon wafers for different microelectronic products in dependence of final chip thickness. The given thickness values reflect current knowledge and experience and correspond to wafers with 150 mm diameter. Dark part of bars represent regions of recommended application.

Handling techniques for thinned wafers	Microelectronic products based on thinned silicon substrates			
	Chip cards, RFId ICs, Power devices	Flexible ICs, thin labels	Vertical System Integration	SOI substrates
Protec...				
Dedicated equipment and tape				
Carrier substrate				
Wafer thickness in μm	800 600 400 100 80 60 40		10 8 6 4	1

Table 2 illustrates the linking of thin wafer handling schemes with expected wafer thickness and possible microelectronic products. Thickness limits for the suggested handling concepts are not finally settled. For a review on the different evaluations concerning the best choice of handling technique see references [13, 14, 15].

Today's primary thin chip products are smart card ICs. Thickness of these silicon devices is in the range of 150 μm to 250 μm. Wafer thinning is done by means of front side protective tapes (grinding tape); backend processing like wafer dicing and pick-and-place of single ICs is done by standard equipment and processes.

Thinnest devices fabricated on ultra thin, single crystalline silicon films are realized on SOI substrates (silicon on insulator). Thickness of silicon films are in the range of 0.2 to 5 μm. SOI substrates are an example of a microelectronic product which have their rigid carrier substrate permanently bonded to the final product chip.

For the intermediate thickness regime, d = 20 .... 80 μm, temporarily bonded carrier wafers are supposed to be an appropriate technique for handling and processing of thinned substrates. Most important requirements for carrier techniques are cost effectiveness of process and equipment and the possibility of multiple use of carriers.

2.3   Dicing of thin wafers

Mechanical stress induced by a standard sawing process at the side wall of a chip may generate an important reliability issue. Therefore dicing technology becomes a challenging task for thin wafers. The situation is comparable to backside grinding, for which sub-surface damage after material abrasion has to be removed by some kind of stress-relief process. For the establishment of thin wafer technology avoiding material damage by any dicing technique is of major interest.

2.3.1   Dicing by means of sawing equipment

Wafer dicing is most commonly performed by means of sawing equipment running with diamond equipped rotating cutting wheels. This process scheme is also applicable for thin wafers. However, it has to be considered that sawing induces micro damage at the sidewall of silicon chips (see fig. 2).

Figure -2. Micrograph of a 40 μm deep dicing line, prepared by means of a wafer saw.

As the wafers become thinner apparent micro cracks may become a serious problem for the reliability of ICs. Mechanical stability of silicon chips is required for safe handling, mounting and packaging of IC products. Hence micro cracks at the chip edge may lead to chip breakage during subsequent pick and place process or pressure supported flip chip mounting.

Furthermore sawing is a sequentially running process. For very small chip sizes, e. g. in the case of small transponder ICs for smart label applications, process time for wafer sawing extends to several hours per wafer.

To overcome the problems of sawing a thin wafer so called "Dicing before Grinding" process (DBG) was developed and proposed by the Japanese companies DISCO Hi-Tec, Toshiba and Lintec Corporation [16]. With DBG concept wafer sawing is performed at front side of device wafer. Instead of completely cutting the wafers, front side grooves are generated. Depth of grooves corresponds to desired final wafer thickness. After this pre-preparation of dicing lines the wafer is transferred from dicing tape to backside grinding tape. Chip separation takes place during a subsequent thinning sequence.

Applying the DBG process reduces material damage induced by sawing and also allows for higher cutting speed. Nevertheless mechanical damage after diamond blade sawing is still present and may influence material properties of ultra thin silicon chips.

2.3.2 Dicing by Thinning by means of reversibly mounted carrier substrates

In the year 2000 an integrated process concept for thinning and die separation was proposed and demonstrated by the authors of this chapter and their working group [17]. The process scheme "Dicing-by-Thinning" ("DbyT") comprises narrow chip grooves at wafer front side and makes use of carrier substrates which are temporarily fixed to the device wafer that undergoes thinning routine (see fig. 3).

Chip separation takes place during backside thinning when finally the front side grooves are opened. At this stage of manufacture device wafers are readily separated into single chips (see fig. 4). Chips can now be picked up from the rigid carrier or transferred as a whole set of ICs to another type of supporting or transporting medium.

Chip grooves may be prepared by wafer sawing or even more favorable by means of dry etching techniques. Groove formation by dry etching practically eliminates mechanical defects at chip edges (see fig. 5).

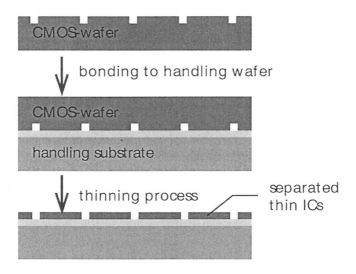

pre-patterning of scribe lines

CMOS-wafer

bonding to handling wafer

CMOS-wafer

handling substrate

thinning process

separated
thin ICs

Figure -3. Schematic sequence of the "Dicing by Thinning" process.

Figure -4. Final state of thin chip separation according to „Dicing-by-Thinning" concept. The micrograph shows separated chips, size 2 x 2 mm^2 having a thickness of 25 μm.

Figure -5. SEM picture of a chip groove prepared by dry etching at front side of device wafer; groove width: 7 μm, depth: 30 μm.

Narrow chip grooves can be fabricated by silicon plasma etching. Layout of trenches are defined by a photo-resist mask. Therefore geometry of chip grooves is not restricted to rectangular shapes. As an example chips may have rounded corners (see fig. 6) or hexagonal shape.

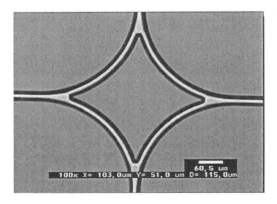

Figure -6. Preparation of front side grooves to generate rounded chip corners for enhanced mechanical stability.

Compared to a sawn dicing lines the width of dry etched grooves can be reduced from 80 μm to 5 μm. This leads to an important economical advantage in the case of very small chips, e. g. transponder ICs. For a chips size of 0,4 x 0,4 mm^2 approximately 40 % of the wafer area can be saved (see fig. 7). This allows for a strong increase of IC products per wafer.

114

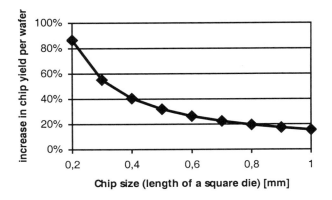

Figure -7. Calculated increase of chip number per wafer for a reduction of the width of dicing lines from 80 μm (wafer saw) to 5 μm (dry etched trenches) as a function of die size.

The various advantages of "Dicing by Thinning" concept can be summarized as follows:

- parallel processing of chip grooves
- savings in silicon area, especially for small die size; resulting in higher chip number per wafer
- ideal chip edges improve fracture strength
- rounded chip corners enhance mechanical toughness
- non-rectangular die layouts are possible
- higher mechanical strength allows for reliable chip handling and mounting

Unfortunately applying an etching process to separate commercial wafers is not straight forward, because scribe lines usually contain test circuits or process control modules, which can not be etched in a simple way. Therefore wafer layout has to be modified to use the "Dicing by Thinning" concept together with trench etching.

3. MATERIAL ANALYSIS OF THINNED SILICON

3.1 Influence of backside treatment to material properties of silicon wafers

3.1.1 Atomic Force Microscopy

Microscopic surface roughness of ground and wet-chemically etched silicon samples were measured. Aim of this test was the observation of the influence of a wet-chemical process to a ground silicon surface. The AFM measurements show (see Fig. 8), that the surface roughness after grinding can be greatly reduced by a spin-etch removal of 10 to 15 μm. Etching more than 20 μm does not lead to a further improvement of surface quality.

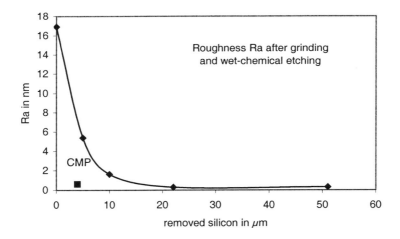

Figure -8. Reduction of surface roughness after grinding: Removal of 15 μm by wet-etching gives similar values as a CMP polish removal of 4 μm (black square).

As an alternative process for stress-relief CMP (Chemical Mechanical Polishing) processing was investigated. It turned out that polishing off 4 μm silicon after grinding generated a similar surface roughness value R_a as wet-chemical removal of approx. 15 μm. This experiment reflects the well known capabilities of CMP processing for high quality surface finishing. However, due to higher removal rate wet-chemical spin-etching allows for higher wafer throughput when compared with polishing.

Observations made by Atomic Force Microscopy do not imply a statement concerning subsurface crystalline quality. Defect free single crystalline silicon is a must for the functionality of integrated circuits. For ultra thin wafers an existing regime of damaged substrate material at backside of a thinned device wafer is getting closer to active electronic layers at the front side. Hence more intense work has to be focused on electrical parameters of thinned silicon.

3.1.2   Electrical properties of silicon after backside thinning

Electrolytic Metal Analysis Technique (ELYMAT) is a standard test method for the detection of metal impurities and crystalline quality of semiconductor substrates [18]. The wafer is inserted into a HF containing bath (see Fig. 9) and it acts as a partition wall between two electrolytic cells. Wafer and cells are connected to opposite electrodes. Minority carriers are induced inside the semiconductor by an optical laser beam. The lifetime of generated carriers is electronically determined. High lifetime values prove good crystal quality.

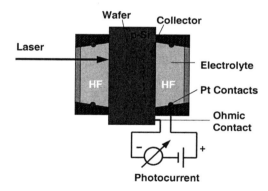

Figure -9. Principle of lifetime measurement (courtesy of GeMeTec, Munich).

Experiments were carried out with differently processed wafers: only ground, CMP polishing after grinding, wet-etching after grinding, etching and CMP after grinding. An untreated silicon wafer (prime quality) was used for reference. The result is illustrated in Fig. 10. Grinding destroys crystal lattice; lifetime of carriers vanishes. Subsequent stress relief processes remove the damaged layer. Initial values of carrier lifetime (see reference wafer in fig. 10) are restored. Also it can be seen from Fig. 10 that the wet-chemically etched backside leads to highest lifetime values. This is a consequence of the very effective wet-chemical surface cleaning which removes any residual metal contaminations from the back side of the thinned wafer.

Due to the experimental setup wafer thickness in this experiment was restricted to thick substrates (d = 500 μm, 150 mm diameter). The backside photo current measurement is a very sensitive method to analyze the different thinning techniques. It also provides a correlation between material defects and electrical functionality of thinned wafers. Therefore it would be an interesting task for the future to modify the experimental setup so that it can be applied also to very thin substrates.

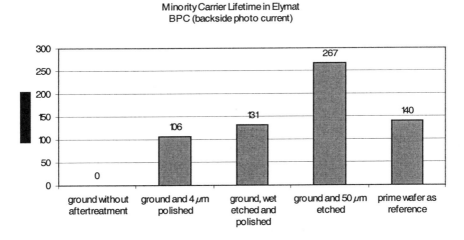

Minority Carrier Lifetime in Elymat
BPC (backside photo current)

Figure -10. Results of carrier lifetime measurements from 500 μm thick wafers after grinding, etching, polishing.

3.2 Mechanical Properties of Thinned Silicon

3.2.1 Investigation of mechanical behavior

Standard methods for testing fracture strength are 3-point breakage experiments and ring ball breakage tests. Later method is applicable to full wafers when using a large diameter support ring. In this case no dicing process influences the mechanical tests. If ring ball tests are applied to diced samples, that are geometrically larger than support ring, the influence of dicing is supposed to be negligible. In the context of wafer backside thinning ring ball tests are recommended to compare different thinning techniques with respect to the resulting breaking strength [19].

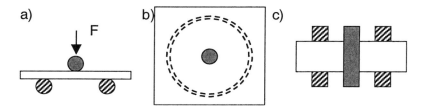

Figure -11. Geometry of bending and breaking tests for silicon test samples: a) side view, b) upper view of ring-ball test, c) upper view of 3-point-flexion test.

With 3-point breakage test, as shown in fig. 11c), the applied dicing technology which is necessary for preparation of test samples has an important influence on the measured values of breaking force.

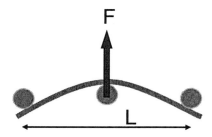

Figure -12. Illustration of the 3-point-flexion test.

An often used relation between breaking force F and mechanical stress S of a test specimen of length L (distance of rods), width B and thickness d is [20] (see also fig. 12):

$$S = 3/2 \cdot F \cdot L / (B \cdot d^2)$$

This formula is valid for deformations which are small compared to the thickness of specimen. As this restriction is not valid for ultra thin and flexible silicon samples, the formula could not be applied to evaluation of experiments. The analysis was carried out using the breaking force values as measured by the testing equipment. Therefore comparison of differently processed thin silicon plates is only possible for specimen of same thickness and geometry.

3.2.2 Experimental

Investigations were performed to compare the fracture strength of thin silicon samples which were processed by different thinning and dicing techniques. A 3-point bending and breaking test apparatus was used. The aim of test series was to determine the depth of damage induced by mechanical grinding and to investigate the possible benefit of a stress relief process.

Experiments were carried out for a sample thickness of 150 μm, 65 μm and 20 μm. Wafer dicing was performed by either sawing after thinning or by using the "Dicing-by-Thinning" concept. In all cases temporarily bonded carrier wafers were used during thinning. First thinning step was grinding (DISCO DFG 850); a 360 mesh grinding wheel was used for rough grinding, a 2000 mesh grinding wheel was used for subsequent fine grinding. Stress relief process after grinding was either performed by wet-chemical spin-etching (SEZ SP203) or CMP polishing (IPEC Westech 371). Three different technological approaches were compared:

a) Sawing after Thinning: Readily thinned wafers were transferred on standard dicing tape; chip separation took place by sawing of the thin wafer.

b) Dicing-by-Thinning using dry etched front side grooves: Chip separation took place during final spin-etch thinning process

c) Dicing-by-Thinning using a wafer saw for groove preparation: In this case front side trenches were prepared by half cut sawing. Influence of stress relief processes spin-etching and CMP polishing were investigated and compared.

When wafers were processed according to DbyT concept (case b and c) opening of front side grooves took place during spin-etching. This means that some parts of sidewalls of chips were over-etched for a few seconds by either slurry media (CMP) or isotropic acid HF/HNO$_3$ (spin-etch).

3.2.3 Results

Fig. 13 and 14 show the results of 3-point-breaking tests for sample thickness 154 μm and 67 μm. The diagrams present the measured values of breaking force in dependence of backside silicon removal after grinding. For both wafer thickness the stress relief processes spin-etching and CMP polishing are compared. Furthermore each diagram in fig. 13 and 14 compares the two dicing technologies: sawing (case a in the previous section) and DbyT with sawn grooves (case c). Measuring points at x-axis value 0 represent the mean breaking force and it's standard deviation of samples which have been thinned by grinding only.

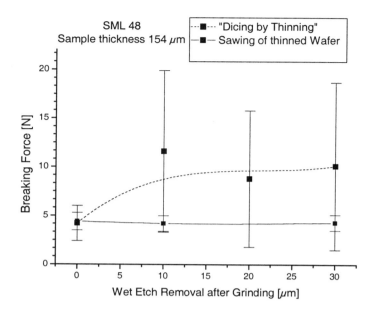

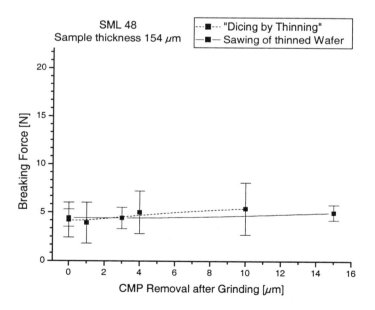

Figure -13. Influence of backside stress relief wet-etching and polishing after grinding to fracture strength of thin silicon test samples having a remaining thickness of 154 μm.

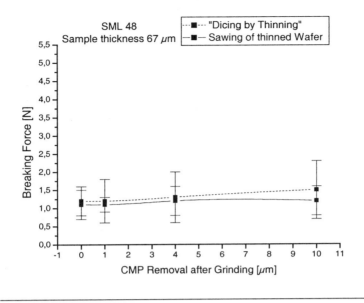

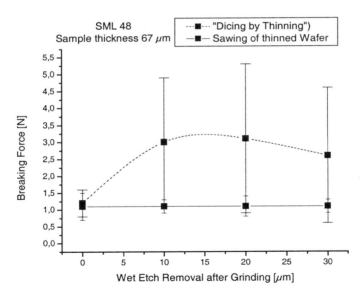

Figure -14. Influence of backside stress relief wet-etching and polishing after grinding to fracture strength of thin silicon test samples having a remaining thickness of 67 μm.

Samples which were sawn after thinning (case a) exhibit no increased breaking force values neither after 30 μm wet-etch stress-relief nor after 10

μm CMP stress-relief. Obviously any benefit of backside stress-relief process got lost during subsequent sawing.

If samples were processed according to "DbyT" concept (case c) wet-etch stress relief process leads to an increase of the mean value of breaking strength by a factor of 2 for 154 μm thick samples and by a factor of 3 for 67 μm thick samples. However, standard deviations of measuring points are dramatically increased also. As it is visible from the diagrams the low force limit of error bars lie in the same range as samples prepared according to case a). The question why error bars are so large leads to an interesting explanation: when front side grooves are opened during backside spin-etching some sidewalls of chips are chemically over-etched. However, due to geometrical conditions of etch flow on the wafer during spin-etching some parts of sidewalls are not attacked by the liquid. So etching conditions for chip sidewalls cannot be controlled in this process step. This finally leads to the found large deviations in breaking behavior.

Similar arguments can be used to explain the measurements for samples which underwent a CMP stress-relief process (case c). Practically no increase of breaking strength is found. Error bars are much smaller than for wet-etch samples because polish process dominantly effects wafer backside. Wet-chemical component of CMP slurry has only poor impact on the sidewall of chips.

Fig 13 and 14 allow a comparison of two sawing techniques: sawing after thinning and pre-cutting of dicing lines at wafer front side. The breaking experiments clearly shows: there is no higher breaking strength for front side grooved wafers. Hence the thesis can be set up, that micro stress induced by sawing is present in both cases and will finally reduce mechanical reliability of thinned ICs.

It's also interesting to have a look on absolute values of breaking force in fig.13 and 14. Sample thickness d differs by a factor of 2. According to previously mentioned formula breaking stress S is inversely proportional to d^2. In the case of zero stress relief (x-axis value 0) the experiments exhibit a factor of approximately 4, as it would be predicted by stress formula.

It can be also seen from the test series that a 67 μm thin sample can reveal higher values of breaking strength than a 150 μm thick sample, if damage at wafer backside and along dicing lines is removed by an adequate stress relief process.

Most promising way to eliminate mechanical stress in thin silicon chips is the preparation of front side grooves by plasma dry etching (case b). Test series of breaking experiments were carried out for 20 μm thin silicon samples [21]. Wet-chemical spin etching was applied as stress-relief process. Material removal by etching was set to 15 μm, 30 μm and 45 μm. Frequency distribution of measured breaking force values are shown in fig. 15.

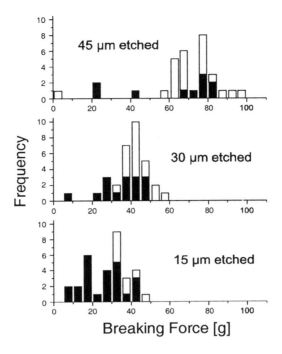

Figure -15. Frequency distribution for a test series of 20 μm thin silicon samples, which were prepared according "DbyT" concept using dry etched front side grooves. White portion of bars indicate samples, that could not be broken.

In this part of test series values of breaking strength for zero stress-relief could not be achieved, because it is practically impossible to get a 20 μm thin wafer by grinding only.

Dependence of breaking force on the amount of stress-relief is given in fig. 16. The behavior is different from results which were obtained in case c).

Firstly standard deviation is much smaller than in the case of sawn chip grooves and secondly values of breaking force are further increased by a higher etch removal. Hence it can be concluded that mechanical behavior of chips which were separated by dry etching is not ruled by dicing stress. Dominant influence is now coming from backside thinning procedure.

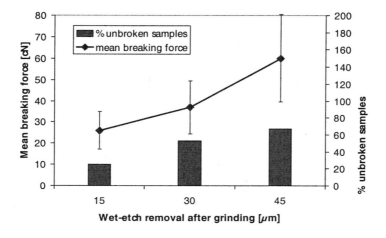

Figure -16. Influence of backside stress relief process to breaking force of thinned chips; silicon samples were prepared according to "DbyT" concept by means of dry etched trenches.

Table 3 correlates the values of breaking force to the applied thinning sequence: Mechanical strength can be doubled when chips are prepared by grinding to 65 μm and etching-off 45 μm instead of grinding to 35 μm and etching-off 15 μm. This is an interesting hint for the thesis that the depth of damage induced by grinding depends on the remaining wafer thickness after grinding.

Table -3. Thinning sequence and results of fracture tests of silicon samples which were prepared by dry etched grooves for chip separation.

Wafer thickness after grinding	Etch removal after grinding	Final thickness	Number of samples	Number of unbroken chips	Mean fracture force in cN	Std. dev. of force [%]
35 μm	15 μm	20 μm	32	8	26	34
50 μm	30 μm	20 μm	32	17	36	27
65 μm	45 μm	20 μm	32	21	59	43

The bars of frequency distribution in fig. 15 are divided into two parts: black parts are related to breaking force values and white parts represent bending force of samples which could not be broken due to their higher flexibility.

Figure -17. Demonstration of the high flexibility of a 20 μm thin silicon sample, bent to a radius of about 2mm.

Fig. 17 impressively demonstrates the high flexibility of a 20 μm thin silicon sample that have been thinned and diced according to "DbyT" concept using dry etched trenches.

3.2.4   Conclusion

Evaluation of breaking experiments lead to the following conclusions:
- Benefit of backside stress-relief process after grinding gets lost, if very thin wafers are diced by sawing after thinning procedure
- Wafer sawing induces mechanical stress in silicon substrates; it does not matter whether sawing was performed before or after backside thinning.
- Highly flexible silicon samples can be achieved, if stress-relief processes are applied to backside of wafers as well as to dicing lines.
- Depth of damage induced by grinding is higher for thinner wafers. Final determination of damaged zone is an important knowledge for the development of an optimized thinning procedure.

3.3   X-Ray Inspection

3.3.1   Reflection Monochromatic Section Topography

Section topography was applied to determine the influence of different thinning techniques to the crystal lattice [22]. Same wafers were analyzed as in

AFM microscopy and lifetime measurements (see previous sections); the wafer thickness is 500 μm.

Measurements were carried out at the European Synchrotron Radiation Facility (ESRF) in Grenoble, France, by Dr. Thilo Baumbach and his coworkers.

Fig. 18 shows a set of monochromatic section topographs of differently treated wafers: a) untreated reference wafer, b) only ground, c) ground and 20 μm wet-etched, d) ground and 4 μm polished.

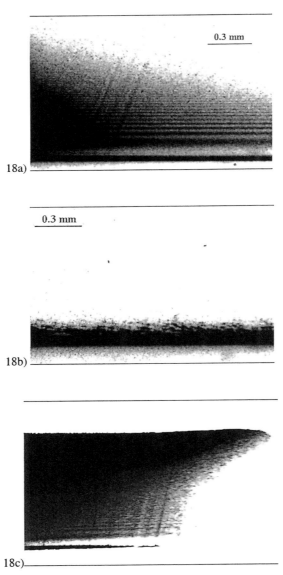

0.3 mm

18a)

0.3 mm

18b)

18c)

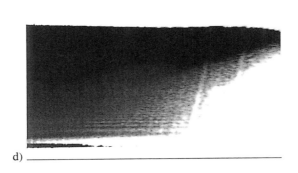

d)

Figure -18. X-ray section topographs of thinned silicon wafers: Pendellösung fringes are used to qualify crystal lattice. a) untreated reference wafer, b) only ground, c) ground and 20 μm wet-etched, d) ground and 4 μm polished.

The pattern of black and white lines are identified as so called Pendellösung fringes. Such pattern originate from multiple diffraction of the x-ray beam inside a perfect crystal lattice. The appearance of Pendellösung fringes for a certain Bragg reflection therefore proves the perfect quality of a semiconductor substrate. Fig. 18a) shows the pattern from an untreated and hence undistorted reference wafer. Grinding destroys the perfect periodicity of the crystal lattice and so the fringe pattern disappears (Fig.18b). After wet-etching off 20 μm backside silicon (Fig. 18c) or polishing off 4 μm (Fig. 18d) the fringes reappear. The number and sharpness of visible Pendellösung fringes (PL) can be used to characterize crystal perfection (see Table 4).

Table -4. Number of Pendellösung fringes (PL) in a x-ray section topograph of silicon wafers (thickness: 500 μm) can be used to compare crystalline quality after different backside treatments.

Backside process	Number of PL fringes
Reference wafer	20
Only ground	0
Ground and 5 μm wet-etch	4
Ground and 10 μm wet-etch	4
Ground and 20 μm wet-etch	7
Ground and 50 μm wet-etch	8
Ground and 4 μm polish	9

The following conclusions can be drawn: Wet-etching or polishing as stress-relief treatment after backgrinding are suitable processes to restore crystalline quality of a thick (500 μm) silicon wafer. However, the initial

128

perfection of a prime quality silicon wafer cannot be restored. Removal of 4 μm silicon by polishing or 20 to 50 μm by wet-etching leads to similar quality. So for thick wafers it can be concluded that the depth of damage induced by grinding can be greatly eliminated by etching or polishing.

Unfortunately it is rather difficult to apply this method to ultra thin samples because larger deformation of the samples would also prevent the appearance of Pendellösung fringes.

3.3.2   White beam projection topography

White beam projection topography was used to get more information on the depth of damaged layer of thin silicon substrates. The samples were thinned by means of carrier substrates for temporarily stabilization during thinning procedure.

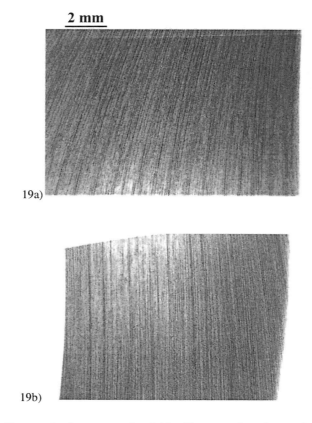

2 mm

19a)

19b)

Figure -19. X-ray projection topographs of thin silicon samples: a) ground to 100 μm; b) ground to 50 μm;

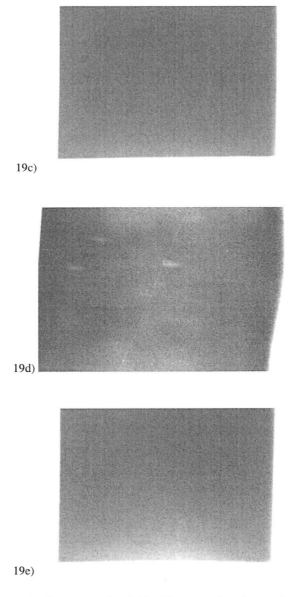

19c)

19d)

19e)

Figure -20. X-ray projection topographs of thin silicon samples: c) ground to 100 μm and etched to 90 μm; d) ground to 50 μm and etched to 40 μm; e) untreated reference wafer.

Finally the thin wafers were cut into pieces of 30 mm x 30 mm size. For the x-ray experiments thin silicon specimens were fixed to a plastic frame holder thus enabling the X-ray beam to penetrate through the sample (Laue transmission geometry) without disturbance of an underlying material.

The topographs clearly show the crystallographic distortions after grinding (Fig. 19 a) and b)). As it can be seen from Fig. 19c) a wet-etch removal of 10 μm eliminates the backside damage in the case that the wafer was ground to 100 μm. If it was ground to 50 μm and subsequently wet-etched for 10 μm distortions in the crystal material and even a micro crack are still visible (Fig. 19d).

The different final quality in topographs 19 c) and 19 d) lead to the thesis that the depth of damage induced by grinding is depending on the thickness after grinding. A final determination of the depth of damage after grinding silicon wafers is yet not possible. Further investigations are necessary to find the minimum thickness after grinding that is tolerable in terms of crystal perfection of ultra thin silicon substrates.

4.   HANDLING OF SINGLE CHIPS

After the thinning and dicing process, the chips are available on a tape. The tape is either putted up on a sawing frame or supported by a rigid carrier e.g. a carrier wafer. The next step in a process flow is pick and place. At the moment, there does not exist an industrially proved procedure. In principle, there are two different ways for this step.

The first one allows the use of the standard die bonding equipment, but it costs the carrier wafer and needs the normal width of sawing groove. The thinned wafer separated in single chips is bonded on the rigid carrier wafer via a thermal release tape. The thermal releasable side of the tape shows to the chip. Depending on the following interconnection process, face up (corresponding to isoplanar interconnection) or down (corresponding to flip chip or flip chip like interconnection), the active side has to show in the right direction. Otherwise the thinned wafer has to be transfer bonded to another carrier. The sandwich consisting of carrier, thermal release tape and thinned chips is bonded to a sawing tape, which is clamped to a sawing frame. Than, the carrier wafer is sawed into single chips of the same size as the active chips (s. fig 7). Sawing of thin wafer and carrier together with the same cut is possible but not recommended because of the strong chipping of the thinned wafer (s. Section 3). Now for pick and place, the chip sandwich can be handled like a standard chip using standard equipment. The eject needle ejects the chip, which is taken by the pick up tool. Then the chip is flipped and placed on the interconnection layer of the final substrate. During the curing or soldering process, the thermal release tape is heated up, so that the carrier chip together with the sticking tape may be removed. Especially in the case of Al-metallization of bond pads, residues of the thermal release layer may be a problem.

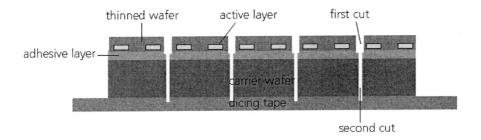

Figure -7. Dicing scheme

If the ultra thin chip is available only on a tape, the normally used eject process with a needle is very critical. This would destroy the chip. In the case of a thermal release tape, the single chip or the whole wafer maybe heated up. The problem of heating up each chip at the moment, where it is picked, is the achievable pick up velocity. The problem at heating up the whole wafer is due to the intrinsic stress, which leads to a rolling up of the chips especially in the case of larger ones. For smaller chips the tape is sticky enough to fix the chips on the tape.

The adhesion forces of a normal sawing tape (e.g. blue tape of Nitto) or UV releasable tape are too high for picking up an ultra thin chip without additional support from the underside. Conceivable is a simultaneous going up of a stamp during picking up.

For an easy pick up process, a "near zero force" tape, whose adhesion is just high enough for chip retaining, would be desirable. But in this case, we have a problem with transfer bonding of the wafer from the tape, which is used for thinning. For lifting off this tape, a certain adhesion force of the pick up carrier tape is needed. At the moment of transfer bond, the adhesion force of the carrier tape must be larger than that of the tape, used for thinning. Therefore a mechanism is needed, which allows the apply of a relatively strong adhesion force during transfer bonding and near zero force to pick up.

5.   ASSEMBLY OF THIN ICS

For electrical interconnection there exist two principles, face up or face down assembly. Face up means, the active side of the IC shows upward, corresponding face down the active side shows downwards. The later one corresponds to flip chip or flip chip like techniques. As the thickness of the IC is very low, wire bonding is possible but generally not recommended for electrical interconnection. The wire loop would be too high. Therefore, for face up assembly, we suggest methods, which we call isoplanar interconnection. This means all types of methods, in which interconnecting conductor lines are led across the chip edges.

In the following sections, the different methods will be explained.

5.1   Face up assembly

Face up assembly corresponds to two separate processes. The first one is the die bond process for the mechanical attaching of the die on the substrate. Electrical interconnection is the second process step independent from the first one.

5.1.1   The Die Bond Process

The die bond process consists of three steps:

First step is the apply of the bond adhesive to the flexible substrate. This could be done by dispensing, stamping or screen/stencil printing.

Second step is the placement of the IC.

Third step is to cure the die bonding adhesive.

With the aim of a low topography of the assembly a critical issue is the accurate dosing of adhesive for die bonding. Thickness of adhesive should be in the range of less then 10 micron. The volume of dosed adhesive depends on the chip size and is about some nano litres e.g. the needed volume of adhesive for a chip with 2x2mm² and an adhesive thickness of $5\mu m$ is only 2nl.

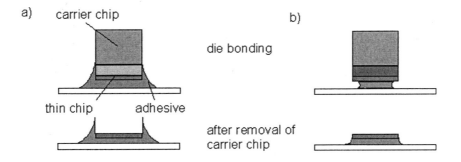

Figure -8 Die bonding of thin chip,
a) Effect of a surplus of adhesive, b) Correct dosing

A surplus of adhesive leads to problems. In the case of using the sandwich consisting of rigid carrier, tape and thinned chip, it leads to the formation of "border mountains". The adhesive wets the lateral faces of the IC [23]. The effect is shown in Figure 8. A topography measurement is shown in Figure 9. In the case of pick and place the thin chip alone without carrier, the pick and place tool and/ or the chip surface will be contaminated by the adhesive.

Too less adhesive leads to incomplete underfill especially at the corners of the IC. Mechanical damages during the electrical connection process may be the result.

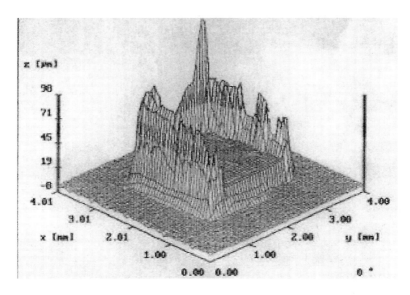

Figure -9. Topography of attached thin IC showing a surplus of adhesive

One possible solution for accurate dosing of low quantities of adhesive is a stamping technique with a very fine needle. Another possibility is printing of the glue. There are some micro dosing systems available on the market. Dosing principle are mostly micro droplets. But their applicability is often restricted to special adhesive materials for instance exclusively UV curable adhesives. Standard materials, which are cured by temperature, will block the outlet in a short time.

There is also a interdependence between dosing method, applied tool, viscosity of adhesive material and surface condition of substrate material. For each combination the process has to be optimised.

Correct underfilling also depends on the pressure of the die bond and furthermore of the correct alignment between adhesive dot and die attachment. For every new combination of adhesive type and substrate material new parameters must be verified. Figure 10 shows a topography of die bonded IC with accurate dosing. Substrate material is a polymer foil and adhesive an epoxy.

134

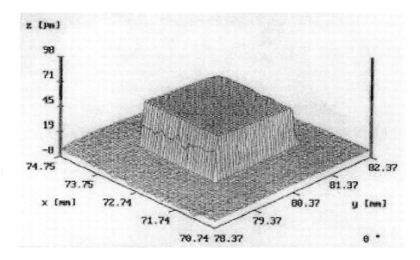

Figure -10. Topography of attached thin IC showing accurate dosing of adhesive

There are endless number of adhesive material producers and they offer a wide variety of different types of adhesives [24]. Very often, the materials are developed for a specific application. Following, the characteristics of some material groups are described.

Cyanacrylates

The advantage of cyanacrylates is the low process temperature that leads to short cycle times. This suited them for foils and paper. On the other side the parameters are not stable during working time.

Epoxies

They guaranty higher adhesion and stability and accurate dosing by stamping or printing. The working parameters are stable for longer time. But on the basis of high process temperature they are only suitable for foils with sufficient temperature stability.

Thermoplastic adhesive

They guaranty rapid bonding but on high temperature. Thermoplastic adhesive also shows high adhesion on foil and paper and low capillary effects.

As mentioned before, the chip placement has to be done with a high alignment accuracy. Otherwise the effects will be the same as a surplus of adhesive material causes.

A short curing time is desirable. Many glues allow a temperature / time regime and often the curing temperature has to be chosen according to the thermal stability of substrate material.

5.1.2 Isoplanar interconnection techniques

The simplest method of isoplanar interconnection is printing or dispensing a silver-filled polymer from the pads on the substrate to the pads on the chip across the edges of the thin chip (s. fig. 11) [25]. Figure 12 shows a cross section of a good looking isoplanar contact. The die is sunk in the adhesive die bond layer and the silver-filled polymer can easily cross the chip edge. To achieve a good contact between silver paste and IC bond pad, a special contact metallization e.g. gold is recommended.

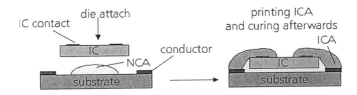

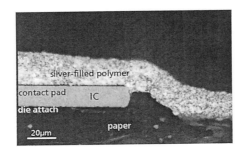

Figure -12. Cross-section of an isoplanar interconnection

This process is cost effective, but is restricted to elements containing only few I/Os. A typical example is shown in figure 13. The shown transponder contains a RFID-IC working at 13,56MHz and a coil with only one turn. Therefore this system is bad adapted to the resonant frequency and therefore the read/write distance is very short. Such a transponder may be used in quasi contact region for application where large read/write distances are undesired e.g. it may be implemented in a banc card.

The production steps are reduced to a minimum: Adhesive applying, placement of the chip and printing of interconnection and coil at once. This kind of face up assembly leads also to the lowest total height of a package, if the IC does not need a back side contact. In this case, the IC is putted directly on the substrate without a metal layer beneath. So, at the chip location the height is only the sum of adhesive layer, IC, interconnection layer and cover foil.

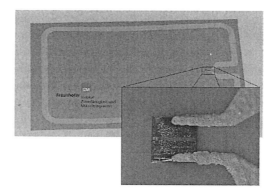

Figure -13. Example of a transponder containing isoplanar interconnection and printed coil

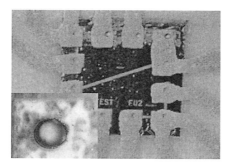

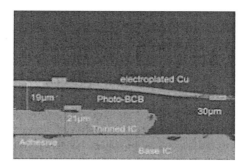

Figure -14. Various isoplanar interconnection methods with chip edge protection
a) Laminated IC with laser drilled vias and isoplanar interconnection b) Metallization and isoplanar interconnection using thin film technology

In some cases, especially if the chip is a bit thicker or for higher pin counts, the chip edge should be covered by an isolation layer. The layer can be a screen printed isolator paste, a spun on polymer or a laminated foil. In the last two cases, the chip pads have to be opened by laser drilling or etching [26, 27].

Some examples are shown in figure 14.

This kind of interconnection may also be used for embedding of chin chips in HDI or multilayer pc boards. To increase the achievable pitch and line resolution, printing techniques must be replaced by thin film techniques like sputtering or evaporation the metal layers and patterning by lithography.

5.2   Face down assembly

Face down assembly corresponds to flip chip (FC) techniques or FC like techniques. The main advantage of this technique is, that chip mounting and electrical interconnection is done simultaneously within one process step. The current mostly applied flip chip technique uses bumped ICs and/or substrates. The interconnection is made by soldering an underfill of an epoxy material is normally needed for stress reduction caused by the different thermal expansion coefficients of silicon and substrate material. After soldering, the gap between chip and substrate is normally $>50\mu$m. Such large gaps are not recommended for chips with a thickness $<30\mu$m.

The applicability of solder technique for ultra thin chips is explained in chapter of C. Kallmayer in this book.

In this section, it will be described the various possibilities using polymeric interconnection. In view of automation and low cost applications, a very important aspect of the diverse bonding methods is the need of pressure during curing or no to reach a secure connection between chip and substrate with a minimum at resistance after curing. If no pressure is needed, the expenditure of the machinery is reduced drastically. Instead of a thermode head for each chip a belt furnace is sufficient for curing.

5.2.1   ACA

Flip chip joining using ACA is increasingly used in electronics industry as less expensive alternative to soldering. Gold coated polymer spheres or massive metal spheres with defined size distribution are dispersed in an adhesive film or paste at a concentration too low to allow electrical conductivity. After mounting ACA particles are squeezed by a defined force between chip and substrate pads and form the electrical contact (s. Fig. 15). During curing, temperature and pressure must be applied simultaneously for freezing the sphere deformation.

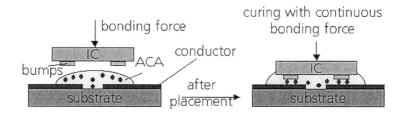

Figure -15. Scheme of flip chip interconnection

In the case of ultra thin chips, the process has to be developed towards minimum bump height and low mounting pressure for two reasons [28]. First, the thickness of the interface layer should be clearly less than the thickness of the IC, second, the volume of adhesive material should be a minimum.

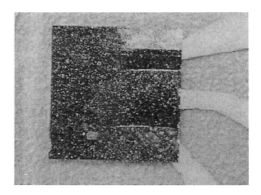

Figure -16. Broken IC caused by incorrect flip chip bonding process

This is necessary, because ACAs show a relative high shrinkage in z-direction during the curing step. This shrinkage and the topography of the underlying substrate induces stress into the chip and in extreme case causes fracture of the chip (s. fig 16).

To reduce bonding time adhesives which tolerate high heating rates are developed. These snap cure materials show a rapid initial hardening to stabilize the assembly.

The demand for more rapid curing leads also to development of adhesives whose polymerisation reaction is started by illumination with visible or ultraviolet light [29]. Besides die-attach, encapsulation and underfill materials also ACA based on these materials are available. Mounting time with applied pressure can be reduced to a few seconds. Final curing can be done at low temperature without pressure.

Even fine pitch applications are possible with this method. For applications with ultra thin chips no bumps are used to maintain flatness and flexibility. Advantages of the ACA flip chip method are:

- low resistance
- state of the art
- high pin counts
- narrow pitches

For making the bonding process easier, the use of an adhesive tape as cover foil has been established conveniently. As shown in fig. 17, the thin chip adheres on the tape, which is oversized in comparison to the chip. During the placing step, the flexible tape tolerates imperfect parallelism between chip and substrate. Additionally, contamination of placing and curing tools is avoided. Fig. 18 shows the cross-section of a 20μm thin FC-bonded chip.

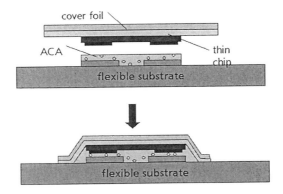

Figure -17. Scheme of flip chip bonding using a cover foil

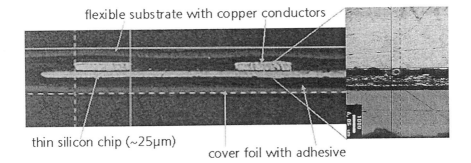

Figure -18. Cross section of a flip chip interconnection

5.2.2 ICA/NCA

A method, which does not need pressure during curing, is the combination of isotropic conductive and non-conductive material [30]. The conductive material is dispensed or screen printed on the contact region and the non-conductive adhesive is dispensed between the contact pads (s. fig. 19). Curing is done without pressure.

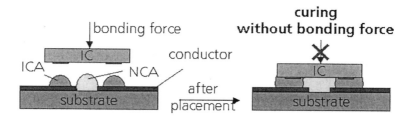

Figure -19. Flip chip bonding using a combination of ICA/NCA

The cross section shows figure 20. Between chip and substrate there is the silver paste of substrate contact and the silver filled adhesive. In the reproduced magnification, they are not distinguishable. The space between the contact regions is filled with non-conductive material. This method is very economical but restricted to large pitches. It is well suited for reel-to-reel processes.

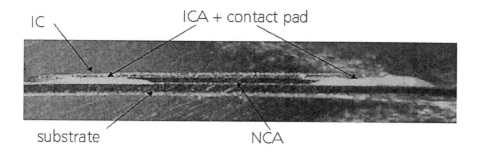

Figure -20. Cross section of a ICA/NCA interconnection

5.2.3 Weak silver filled epoxy

Common ACAs are expensive materials, because of manufacturing of the gold metallized particles with precise size definition. For low cost and mass production products, containing only few contacts like passive transponders with only two, a further step to simplification of the process is possible. It is based on a material, whose specific resistivity lies in the range of about 0.1 to 10 Ohm centimeter. Using geometry effect, a very short resistor with large cross section defines the contact resistance and a long resistor with small cross section defines the isolation resistance. The example with a chip length of 1,5 mm and a contact pad size of axa/3 adhesive thickness of $10\mu m$ led to a resistance ratio of 2500 (s. fig 21). The thinner the adhesive layer the larger is the resistance ratio!

The mechanical mounting and electrical interconnection can be achieved by silver flakes used in common silver-filled conductive adhesives. To reach anisotropic conductivity the concentration of silver particles has to be reduced to an appropriate concentration. This can be achieved by diluting an isotropic conductive adhesive (ICA) like silver-filled paste with the epoxy compound until silver particles have no connection between themselves. In this case the silver paste is able to transmit electricity only in z-Axis and not in the mounting plane. To minimize the gap between chip and substrate, the adhesive layer should be applied as thin as possible and a relative high pressure should be applied during placing.

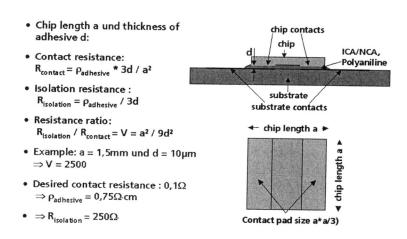

Figure -21. Flip chip-contacts with ICA using geometry effect

The method can be characterized by the following properties:

- cheap silver paste
- higher pressure during chip placing
- no pressure during curing
- large pitches

5.2.4   Intrinsically conductive polymers

Intrinsically conductive polymers are new materials which find increasing applications in electronic industry. Polyaniline is one of the most known organic conductive materials. Formulations of polyaniline can also used for assembly of transponder ICs, which have only few contacts. Resistivity of these materials is rather high (5 Ohm·cm).

For low contact resistance and high isolation between two contacts also the geometric effect is used. By reducing the gap between and chip the lower the contact resistance gets lower and isolation resistance between contacts increases.

With suitable geometrical factors of pad size, pad distance and bonding gap an acceptable assembly for transponders can be achieved. Today this method is not sufficiently investigated, but it may attract interest for future polymer electronics.

5.3   Reliability investigations

For product qualification, reliability has to be proven. At the moment, for thin electronic labels and other applications containing ultra thin chips on flex materials do not exist a standard test procedure. One can lean to test standards for chip cards. But mechanical load is allowed to be much higher than for chip cards.

To evaluate the different mounting techniques two different test samples have been fabricated. The first one consists of a test chip whose conductor pattern forms a short circuit between two substrate pads. With this assembly the resistance of the conductor path including two interconnections can be measured. Only a 2-pole configuration has been used and contact resistance could not be separated from line resistance.

As a second test sample RFId-ICs are mounted on conductor coils and the operation of these smart labels has been checked by wireless read/write operations.

In both cases, the chip pad metallization were gold [31].

5.3.1   Thermal cycling

Thermal cycling between –40°C and 85°C, 15 min at each temperature has been performed to investigate thermo-mechanical reliability of the different assembly methods.

An overview of the results together with main processing parameters is given in table Y1 and Y2

Table -1 Results of thermal cycling test, test vehicle RFID transponder

Interconnection technique	coil material	adhesive	labels tested	after 500 cycles	after 1000 cycles
Flip chip	etched copper	1	9	no failure	no failure
Flip chip	etched copper	2	10	no failure	no failure
Flip chip	etched copper	3	10	no failure	1 defect
Flip chip	printed silver polymer	1	12	no failure	no failure
Flip chip	printed silver polymer	2	9	no failure	no failure
Flip chip	printed silver polymer	3	10	no failure	no failure
Isolplanar	etched copper		8	no failure	no failure
Isolplanar	printed silver polymer		6	no failure	no failure

Table -2. Results of thermal cycling test, test vehicle RFID Transponder, bond method face down, etched copper metallization

Adhesive	tested labels	after 1300 cycles
ACA Type A	10	10/10 passed
ACA Type B	10	10/10 passed
Light activated ACA	10	9/10 passed
Combination ICA/NCA	10	10/10 passed
Dispersed silver	10	10/10 passed
Polyaniline	10	10/10 passed

As these first tests show, thermal cycling seems not to be a problem using any one of the assembly methods.

5.3.2   Humidity

In contrast to thermal cycling, humidity storage tests show differences between the diverse assembly methods and materials. In tab. Y.3 some results are depicted. The test conditions were 60°C/ 93% rel. humidity

Table -3. Results of humidity storage, test vehicle RFID Transponder, bond method face down, etched copper metallization

Adhesive	tested labels	after 144 hours	after 288 hours
ACA Type A	10	9/10 passed	9/9 passed
ACA Type B	10	2/10 passed	2/2 passed
Light activated ACA	10	8/10 passed	8/8 passed
Combination ICA/NCA	10	10/10 passed	10/10 passed
Dispersed silver	10	10/10 passed	10/10 passed
Polyaniline	10	8/10 passed	8/8 passed

Especially ACA type B produced a lot of early failures. Therefore a second test was carried out. It confirmed the first one with one exception, ACA B has not shown the bad result of first test. (s. tab. Y3). For the first test, we assume that in the case of ACA B pressure conditions were not correct during curing. But the results show a tendency of early failure for all ACA adhesives. Best results with no failure in both test cycles were received with the ICA/NCA combination and the low filled silver adhesive.

Table -4. Results of humidity storage (test vehicle s. table 2)

Adhesive	tested labels	after 144 hours	after 288 hours
ACA Type A	15	14/15 passed	14/14 passed
ACA Type B	15	14/15 passed	14/14 passed
Light activated ACA	15	13/15 passed	13/13 passed
Combination ICA/NCA	15	15/15 passed	15/15 passed
Dispersed silver	15	15/15 passed	15/15 passed
Polyaniline	15	13/15 passed	13/13 passed

5.3.3   Mechanical bending

Mechanical bending of ultra thin silicon should be not a problem, as the high flexibility demonstrate, which is described in chapter Y1.1(CL). A closer examination is needed for the chip interconnections. Potential problems can be minimized by putting the chip in the neutral axis of a package. In our example of assembly on flexible substrates, the IC should be embedded between two foils. As before, the test vehicles were RFID labels containing $25\mu m$ thick transponder ICs. The labels were bent over various radii. The minimum radius was less than 3mm, at which the chip was broken or the interconnection opened.

6.   OUTLOOK

Current trends in the development of electronics systems show, that the provision of extremely thin components and semiconductors plays a decisive role in the steadily progressing development of highly integrated systems. A new generation of thin flexible electronic systems arises. One of the most essential branches of industry will world widely be the information and communication technology within the next years. The application of new information and communication equipment will influence our everyday life to an increasing extend. Examples for this are flexible communication systems and equipment which are used in local networks like wearable electronics will be based on highly integrated functions and on flexible electronic equipment correspondingly wearable electronics will be a part of the body area network node . These systems also will contain components like flexible batteries and displays besides the active electronic components.

For such thin, flexible systems, reel to reel processes may be the choice as a very economic production method and will play an important part in the future, especially for low cost mass production.

An example for high-volume mass production can be found in the paper printing industry, where paper is printed with several hundreds of meters per minute in a reel to reel or flexible sheet based production line. New tools and methods will have to be developed. Also new materials for substrates, interconnects and active devices are certainly needed. But at present already standard processes are successfully adapted or converted and extended into a reel to reel manufacturing line as e.g. screen printing or lamination processes which can run continuous and very fast.

Other process steps like chip and SMD assembly, electrical test using test needles or wireless, high resolution patterning using lithography work by stop and go in a reel to reel line. But this is state of the art. In future, there also continuously running methods will be introduced . The usage of presently available materials, tools, equipment and processes offers already a wide spectrum of new applications in order to improve present products or to produce them at a lower price. Many near future products will only be produced if suitable reel to reel processes are available.

Application of reel to reel production offers additionally to combine standard electronic assembly of thin chips and components on flexible substrates with polymer electronics.

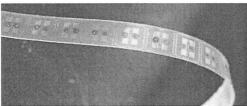

Figure -22. Low-cost Polymertransistors (organic field effect transistors OFET) with dispensed organic semiconductor poly-3-hexathiophene (P3HT) on flexible substrate. pad-size 3x3mm

The age of polytronics (**poly**mer+elec**tronics**) has begun. It is not primarily a replacement for existing electronic technologies, but opens up the prospect of completely new applications that combine the features of transistor, LED, detector and interconnect devices with the freedom of design, flexibility and low cost of plastics combined with highly integrated conventional ICs. In the scope of these new findings it seems possible that polytronics may solve present systems integration problems and add new functionality to microelectronic circuits and systems. Polytronics creates a new and very promising technological area with new applications and products. Examples are:

- Full polymer transponders RFID
- Printable tags
- Flexible systems,
- Disposable electronics,
- Body area networks, smart clothing wearable computing
- Pervasive computing and communication systems
- polymer based photovoltaics, detectors, imaging applications, displays, illumination systems
- Disposable low cost sensors (bio-chemical) actuators.
- Disposable low cost optical and electronical readable memories

7.   REFERENCES

[1]   E. Jung et al.: "Ultra thin chips for miniaturized products", Proc. Micro System Technologies 2001, Düsseldorf 3/2001, ISBN 3-8007-2601-7, 443-448

[2]   P. Ramm et al.: "Three dimensional metallization for vertically integrated circuits", Micoelectronic Enineering 37/38 (1997), 39-47

[3]   R. Wieland et al.: "InterChip Via Technology for Vertical System Integration", Proc. Workshop Thin Semiconductor Devices Manufacturing and Applications, München 12/2001

[4]   M. Feil et al.: "Interconnection techniques for ultra thin ICs and MEMS elements", Proc. Micro System Technologies 2001, Düsseldorf 3/2001, ISBN 3-8007-2601-7, 437-442

[5]   M. Feil: "Technologie-Konzept gekrümmt aufgebauter Photodetektoren", Proc. Optik und Optronik in der Wehrtechnik I, Meppen 9/2001, ISBN 3-965938-00-4, 6/21-30

[6]   C.K. Harmon et al.: "Intelligent labels, the intelligent choice?", Proc. Frontline Solutions Europe 2000, Frankfurt/Main 11/2000, Section W-02

[7]   A. Plettner: "Elements of RFID systems integration", Proc. Workshop Thin Semiconductor Devices Manufacturing and Applications, München 12/2001

[8]   G. Schelhove: "From 13.56MHz-RFID-labels to self-adhesive smart label", Proc. Workshop Thin Semiconductor Devices Manufacturing and Applications, München 12/2001

[9]   C. McHatton, C. M. Gumbert, "Eliminating backgrind defects with wet chemical etching", Solid State Technology, November, 1998, p. 85-90.

[10] B. Holz, K. Kobayashi, "The Slimming Club", European Semiconductor, April 2001

[11] S. Savastiouk, O. Siniaguine, M. Hammond, "Atmospheric downstream plasma", European Semiconductor, June 1998.

[12] C. Krutzler, "Thinning and dicing by dry etch", proceedings of workshop "Thin Semiconductor Devices – Manufacturing and Applications", December 3-4, 2001, Munich.

[13] W. J. Kröninger, F. Hecht, G. Lang, F. Mariani, S. Geyer, L. Schneider: „Time for Change in Pre-Assembly ? The Challenge of Thin Chips", Electronic Components and Technology Conference, ECTC, 2001, Orlando, Fl.

[14] H. Pristauz, W. Reinert, "Handling Concepts for Ultra Thin Wafers", 13th European Microelectronics and Packaging Conference & Exhibition, May 30 – June 1, 2001, Strasbourg, France.

[15] C. Landesberger, S. Scherbaum; G. Schwinn; H. Spöhrle: "New Process Scheme for Wafer Thinning and Stress-free Separation of Ultra Thin ICs"; Microsystems Technologies 2001, Düsseldorf.

[16] K. Priewasser, "Dicing before grinding (DBG) with stress relief", proceedings of workshop "Thin Semiconductor Devices – Manufacturing and Applications", December 3-4, 2001, Munich.

[17] M. Feil, C. Landesberger, A. Klumpp, E. Hacker, „Method of subdividing a wafer", patent application WO 01/03180 A1.

[18] P. Eichinger: "New Developments of the ELYMAT Technique", Recombination Lifetime Measurements in Silicon, ASTM STP 1340, D. C. Gupta, F. R. Bacher, W. M. Hughes, Eds., American Soc. for Testing and Materials, 1998, pp 101-111.

[19] W. Kröninger, G. Lang, F. Hecht, T. Stache, A. Hansen, L. Schneider, „The stability of silicon chips", proceedings of workshop "Thin Semiconductor Devices – Manufacturing and Applications", December 3-4, 2001, Munich.

[20] Walter D. Pilkey, "Formulas for Stress, Strain and Structural Matrices", John Wiley & Sons, New York, chapter 3.12, 1994.

[21] C. Landesberger, G. Klink; G. Schwinn; R. Aschenbrenner: "New Dicing and Thinning Concept Improves Mechanical Reliability of Ultra Thin Silicon"; International Symposium and Exhibition on Advanced Packaging Materials, March 2001, Braselton, Georgia.

[22] C. Landesberger, D. Bollmann, G. Klink, K. Bock, H. Reichl, "Reliability of Thinned Silicon ICs", Proceedings Materials Week 2001, Int. Congress on Advanced Materials and Processes, Oct. 2001, Munich, Germany

[23] G. Klink et al.: "Ultra thin Ics open new dimensions for Microelectronic systems", Advancing Microelectronics, Vol. 27, No. 4, July/August 2000, 23-25

[24] C. Adler et al.: "Assembly of ultra thin and flexible ICs", Proc. Adhesives in Electronics, Helsinki, 6/2000, ISBN 0-7803-6460-0, 20-23

[25] M. Feil, K. Haberger, A. Plettner: "Anwendungen und Technologie extrem dünner Ics", Proc. Systemintegration in der Mikroelektronik, Nürnberg 5/1999, ISBN 3-8007-2456-1, 191-198

[26] M. Töpper et al.: "Thin chip integration – A novel technique for manufacturing three dimensional IC-packages", Proc. 33rd Int. Symposium on Microelectronics and Packaging, IMAPS, Boston 2000, 208ff

[27] G. Klink et al.: "Innovative packaging concepts for ultra thin integrated circuits", Proc. 51st. Electronic Components and Technology Conference, Lake Buena Vista 6/2001, ISBN 0-7803-7040-6 (CD)

[28] M. Feil et al.: "Interconnection techniques for ultra thin ICs and MEMS elements", Proc. Micro System Technologies 2001, Düsseldorf 3/2001, ISBN 3-8007-2601-7, 437-442

[29] Delo "Technische Information, Delo-Katiobond VE 12074"

[30] M. König, G. Klink, M. Feil: "Fast flip chip assembly for reel to reel manufacturing", Proc. Polytronic 2001, Potsdam 10/2001, ISBN 0-7803-7220-4, 319-323

[31] M. Feil, G. Klink, M. König: "Ultra thin chip mounting for flexible systeme", Proc. Workshop Thin Semiconductor Devices Manufacturing and Applications, München 12/2001

Chapter 6

Flexible, Microarray Interconnection Technique Applied to Biomedical Microdevices

Joerg-Uwe MEYER

Martin SCHUETTLER

Oliver SCHOLZ

Werner HABERER

Thomas STIEGLITZ

Fraunhofer Institute for Biomedical Engineering (Fraunhofer IBMT)

1. Introduction

Miniaturization of electronic components, sensors and actuators as well as advances in high density interconnects and hybrid integration provide a challenging technology base to the medical industry for developing a new bread of medical devices that interface directly to cells and tissues. Examples of advanced microimplants and instruments include systems for controlled drug delivery, cell sensing, tissue inspection, body fluid handling and neural stimulation. In this chapter we describe devices which are characterized by an array of multiple fine pitched microelectrodes for which novel interconnects, assembly techniques, and biocompatible packaging procedures have been developed.

The active microimplants demand a high degree of device miniaturization and functional density without compromising on design flexibility and biocompatibility requirements. The stringent requirements for integrating microelectronic components to microstructures that are in direct contact of body tissue led us to the development of small pitch interconnects with a pitch smaller than 100 μm. This chapter describes our flex-tape based interconnection technology – termed "Microflex Interconnect, MFI – and the utilization of MFI in medical devices. In the beginning of this chapter, an overview is given on existing small pitch interconnects and on the montage of bare chips on rigid and flexible substrates. The shortcomings of available technologies are discussed under the light of applying available standard technologies to biomedical devices and implants.

1.1 Motivation to develop flexible, space-economic interconnect

Active biomedical implants employ electronic and technical systems to restore physiological functions that are lost or impaired either by disease or injuries. Smallest footprint, high functionality, highest reliability and biocompatibility are one of the most important requirements for active medical implants. The smaller the device the less invasive is the procedure of inspection or implantation. Furthermore, devices need to be packaged to have a biocompatible tissue interface and to withstand the biodegradation caused by body reactions. Packaging consumes most of the space in biomedical implants. Established biocompatible materials for housing electronics are titanium cases – as used in pacemakers -, glass and inert ceramics. This kind of packaging is rigid with dimensions exceeding sizes of several centimeters. Bendable structures like electrode wires and lead outs are insulated and wrapped in silicone rubber (PDMS), polyurethane and other polymer materials which have been extensively tested for their biocompatible material properties. The development of flexible polymer carriers for holding and interconnecting chips and miniaturized components offer the possibility to develop microelectronic and microoptical systems which are in direct contact with delicate soft tissues and biological structures. Instead of standard housed IC components bare or 'naked' silicon chips and dice are used for hybrid integration to minimize size. The combination of mould array process (MAP) and ball grid arrays (BGA) yield package sizes from 1.6 to 1.3 mm [1]. Chip scale packages (CSP) combine Flip-chip (FC) technology with surface mounting technology (SMT) and BGA [2]. Wafer level CSPs (CSP-WL) lead to smallest package footprints since no additional interposers, lead frames and polymer layers are needed [3]. The process involves a six step redistribution and bumping procedure to prepare the CSP chip for SMT. The CSP chips can be mounted on rigid FR4 boards [2] as well as on rigid-flex [3] or flexible substrates and carriers [4;5].

1.2 Objectives for designing the Microflex Interconnects technology

This paper introduces a 'MicroFlex Interconnection (MFI)' technology. The MFI technology employs photolithographic processes for the fabrication of an ultra-thin (5-10 μm) foil or flat ribbon cable. Conducting structures in the foil interconnect between conventional Aluminum pads of single silicon dice and other microelectronic components in three dimensions requiring a minimum of space. No additional solder layers or bumps are needed to prepare the chip. The chip is mounted on the flexible substrate in a rivet-like approach employing thermosonically bonded gold studs. Figure 1 sketches the MFI technology interconnecting a bare biochip with microwell to a PCB structure that serves as a connector.

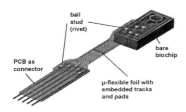

Fig. 1. Schematic of a micropatterned flexible foil interconnecting a bare biochip with a PCB. Ball studs provide for a rivet-like mechanical and electrical interconnection between the open pads of the foil and pads on the bare chip and on PCB.

The footprint of the interconnect is as small as the size of the chip. MFI allows high-density interconnects with center to center pad distances less than 100 μm. The applied materials and material compositions have been extensively tested for biocompatibility since the method was originally invented for neural implants. The MFI technology has served as a high density interconnection and packaging technology to integrate microelectrode arrays and silicon chips for telemetry and nerve stimulation on a single, highly flexible substrate with smallest footprint. Examples of applied MFI technology in neural implants will be given in this text.. The polyimide based MFI structures allow soldering of surface mount devices (SMDs) and other directly onto the flexible carrier. In the following, we discuss material choices of the MFI technology to comply with biocompatibility issues of medical devices and we describe the microfabrication of MFI carriers, the assembly of microcomponets, reliability tests of the interconnects, and the design and performance of MFI based medical devices.

2. Material and Methods

2.1 Identification of appropriate flexible material

Biomedical implants have to fulfill different requirements for a long-term implantation. They have to be stable in the physiologic environment, i.e. no degeneration should occur. All parts of an implant have to be non toxic for the cells or have to be encapsulated with a non toxic material that serves as a diffusion barrier for these substances. The devices have to be designed with a "smooth" geometry to prevent induced trauma of tissue and nerves by sharp edges and corners. Additionally, the material itself should not be brittle or heavy weighted. The implant substrate should be flexible to conform to the natural soft tissue structure. A retinal implant is a good example. The shape of the implant structure has to adopt to the concave shape of the inner eye. Among other polymers, polyimide (PI) was identified as a potential candidate which serves as substrate and carrier for microelectrodes, silicon chips and embedded cable strands. Certain polyimides have proven to be biocompatible when interacting with blood [6]. Polyimide has also been used as an insulator on flexible 125 μm thin Kapton carriers with vacuum evaporated thin film multielectrode arrays [7]. The polyimide was patterned photolithographicly and etched with photoresist developer. The whole microelectrode structure had been applied to record cardiac signals inside the wall of the heart. Various polyimide types are available on the market. We performed cytotoxicity tests according to ISO 10993 to confirm the biocompatible performance of the polyimides. Most of the examined polyimides turned out favorable to be used as biocompatible implant material. Inflammatory reactions of Polyimide in cochlea implants were modest, when present [8].We have chosen the polyimide PYRALIN PI 2611 (HD Microsystems). PI 2611 was particularly selected because of its relatively low water uptake and its thermal expansion coefficient near to the one of silicon nitride (Si_3N_4). The PI 2611 exhibits excellent insulation characteristics and dielectric strength at a low density as well as high material flexibility. In Table 1, the material and physical properties of PI 2611 are compared the ones of photo-definable benzocyclobutene (BCB Cyclotene 3022, Dow Chemical) [9] and to silicon nitride, generated by plasma enhanced chemical vapor deposition (PECVD) .

	Water Absorption @ 85 °C @ 50% RH	Coefficient of Thermal Expansion (CTE)	Glass Transition Temperature	Dielectric Constant @ 50% RH	Volume Resistivity	Breakdown Voltage	Young's Modulus	Tensile Strength
	[%]	[ppm·K^{-1}]	[°C]		$\Omega \cdot$ cm	[V·cm^{-1}]	[GPa]	[MPa]
PI 2611	< 0.1	3	> 400	2.9 * * @ 1 kHz	> 10^{16}	> 2·10^{6}	8.5	350
BCB (Cyclotene 3022)	0.2	52	> 350	2.7 * * @ 1 MHz	> 10^{19}	3·10^{6}	2.0	85
Si$_3$N$_4$	0	3	> 800 * * creep	2.7 * * @ 1 kHz	10^{16}	1·10^{7}	385	14000

Table 1. Material properties compared between Polyimide PI 2611, photosensitive benzocyclobutene BCB and silicon nitrite Si$_3$N$_4$.

PI 2611 is processed with standard cleanroom equipment for microelectronics. It is patterned using reactive ion etching (RIE). The processing technology of PI 2611 is described in the following section.

2.2 Microfabrication of polyimide substrates with embedded metal films

We have developed a process technology for manufacturing ultra-thin highly flexible substrates on which metal microelectrodes, conducting tracks, and interconnection pads are deposited. The final PI design provides for interconnecting silicon chips and SMDs directly on the thin PI film when using the MFI technology. The thin PI ribbon films and the MFI are manufactured with standard thin film and micromachine technologies. To reduce interfaces and expensive packaging techniques the electrodes and substrates were designed using the same material in a single process. The multi-layer process for integrating metal tracks, micro vias and MicroFlex interconnections is illustrated in Figure 2.

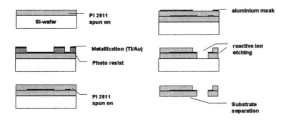

Fig. 2. Photolithographic process of patterning PI films and of the deposition of thin film metal layers on top of the film.

At first, a layer of polyimide is spun onto a silicon wafer. The metallization (Ti/Au,Pt,Ir) for interconnection pads, connecting lines, and electrode sites was deposited in a subsequent step using thin film sputtering. Structures were patterned applying a lift-off technique. After metallization, a second PI layer for insulation was spun onto the wafer. RIE was used for patterning vias (10 μm) into the second PI layer. Subsequently, the PI substrates with the embedded metallization layers were stripped from the wafer. The single devices were lifted manually from the wafer with tweezers. The PI ribbon substrates exhibited a total thickness not exceeding 10 μm. The substrates were highly flexible and bendable.

2.3   Montage of chips and SMDs on flexible substrates

When devising active neural implants it is mandatory to integrate passive and active electronic components close to the microelectrodes. We developed a new assembly method to interconnect ICs and surface mount devices (SMDs) with our highly integrated flexible ribbon substrates achieving smallest packaging dimensions. We termed this technique "MicroFlex Interconnection" (MFI). One of the key characteristics of the MFI is the introduction of connection pads that have a central via. The via is used to place a stud on top of the metal pad, which connects through the via of the ribbon cable to the bottom chip. The 'rivet-like' interconnection is performed utilizing a common thermosonic ball bumping process. A gold ball is bonded onto a metal pad by applying force, temperature, and ultrasonic energy. . A graphical three-dimensional representation of the MFI process is given in Figure 3. The gold ball acts as a stud or metal 'rivet' that electrically and mechanically interconnects the metallized PI carrier with the chip or with other microcomponents or substrates underneath.

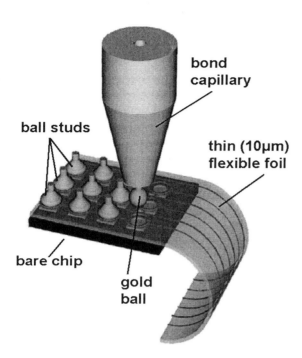

ball studs

bond
capillary

thin (10µm)
flexible foil

bare chip

gold
ball

Fig. 3. Illustration of the "Microflex Interconnect (MFI)" technology. A bond capillary forms ball studs that act like rivets when thermosonically bonded through the flexible foil onto the chip pad.

The different process steps are illustrated in Figure 4. After aligning the PI ribbon pads and vias with the metal pads of the bottom structure a ball-wedge process is initiated. The capillary is lifted and the wire is separated from the bond. Since the via holes of the PI structure are slightly smaller than the gold stud, but large enough to achieve a gold ball bond through the via hole onto the substrate pad, a robust mechanical and electrical interconnection is achieved. Ball-wedge bonding is performed at a temperature of 140 °C, at 60 kHz, and a bond force of 50 cN.

156

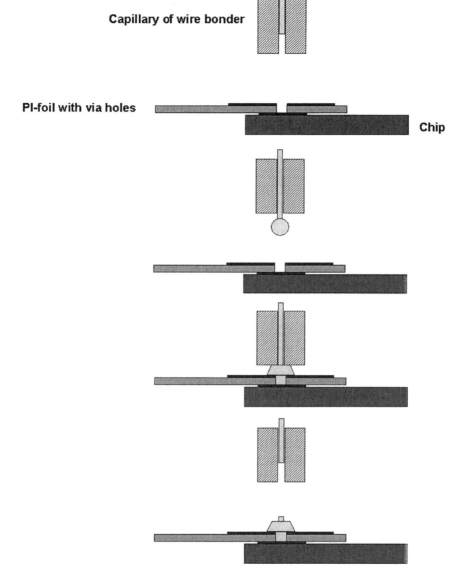

Fig. 4. The "Microflex Interconnect (MFI)" technique as a 4 step process: 1) aligning foil and polyimide (PI) film, 2) ball forming with bond capillary, 3) thermosonic bonding on chip through the PI foil, 4) withdrawal of wire leaving rivet-like ball stud.

Typical design parameters have been pads with a size of 100x100 μm², and a centered via hole with a diameter of 50 μm. For ball bonding, gold wires were used with a diameter of 35 μm. Also wedge bonding was applied. A wedge was produced with gold wires as small as 13 μm in diameter. Figure 5 shows a metallurgic photo through the ball stud which connects the metal pads of the polyimide flex-cable with the bottom chip. A tight connection of the ball stud with the polyimide foil and the bottom substrate is visible.

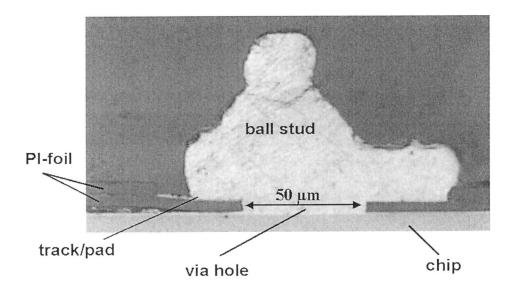

Fig. 5. Metallurgic micrograph of ball stud. The ball stud interconnects the pad of the foil through a via hole with the chip underneath.

With a bending curvature of less than 0.2 mm, the polyimide based μ-flexed cable can be utilized to interconnect between double stacked chips, pair-wise separated by a heat sink or aluminum cast. Figure 6 illustrates the one-to-one array interconnection of two pairs of chips, where the top pair is interconnected with a ball-stud / ball-stud junction to lead out the cable. The bottom pair is interconnected employing a bull-stud / chip pad junction.

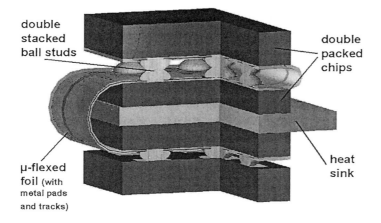

double
stacked
ball studs

double
packed
chips

μ-flexed
foil (with
metal pads
and tracks)

heat
sink

Fig. 6. Illustration of chip stacking with the MFI technology.

The high density packing of 5 chip-pairs is shown in Figure 7. A protruding latch is part of the flex-cable providing an interconnect to a standard plug. The left photo shows ten stacked and interconnected chips from which the plug connector extends in the middle. In the right photo, the five paired chips are placed on an aluminum block with open spaces to separate each pair. The aluminum block is serving as a heat sink and as a cast to house the interconnected chips.

FIG. 7. MICROGRAPH OF MFI INTERCONNECTED CHIP STACKS. LEFT: DOUBLE PACKED CHIPS WITH INTEGRATED CONNECTOR. RIGHT: DOUBLE PACKED CHIPS SEPARATED BY ALUMINUM BLOCKS. BAR CORRESPONDS TO 5 MM

3. *Reliability of Microflex Interconnects*

Test structures were designed for investigating possible resistance changes of MFI / ball stud interconnects under various conditions. In a previous design, contact resistance was measured in a four-wire measurement configuration [11, 12]. Two ball studs were placed through the centered 50 μm via of two MFI pads onto an aluminum substrate. Each of the MFI pads was connected to two large gold pads (300 nm thin, 3,5 x 3, 5 mm[2)], 300 nm layer). The large gold pads were insulated against the aluminium block by the 5 μm polyimide layer of the flex foil. A 1.5 mm thick aluminum substrate was used to keep the resistance of the substrate as low as possible for not interfering with the resistance measured for the bull studs. Aluminum was chosen since it serves commonly as material for contact pads on bare chips. The exact measurement configuration has been shown previously [12]. Interconnect contact resistance to bare chips and SMDs are tested in a standard LCCC 8411 carrier configuration (Figure 8). A microflexed foil was designed to connect to the carrier. One large chip, 5 small chips, and for SMD resistors were connected to the μ-flexed foil. The SMDs were soldered directly to the gold pads of the foil. Different via diameters (50, 60, 80, 100 μm) were fabricated into the foil to investigate potential performance differences.

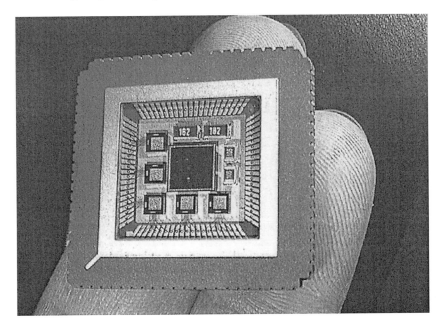

Fig. 8. Test carrier LCCC 8411 holding 6 test chips and 4 SMD resistors on a 20 μm thin polyimide substrate.

3.1 Experimental set-up and procedures

Various tests were performed according to MIL standard 883 to investigate durability and reliability of the MFI interconnects. The contact resistance was measured before and after exposure to the testing procedure. The impedance was measured with an HP4284A LCR-meter (HP4284A) at a current of 20 mA and a frequency of 100 Hz from which the contact resistance was calculated using the analog model described elsewhere [12]. Repeated measurements were taken and averaged.

Thermal testing comprised temperature cycling, thermal shock, biased temperature and humidity conditions, and high temperature exposure. The test devices were placed in a climatic chamber (Heraeus HC 7005) with air cycles between –40 °C and 150°C. A heating and cooling rate of 2 K/min was applied. The final temperatures were held for 5 min. In total, the test devices had been exposed to 84 cycles. Effects of biased temperature and ambient air were tested in the climatic chamber at a temperature of 70 °C and a relative humidity of 95% for 76 hours. The test devices were thermally shocked, alternating between liquid nitrogen (-180 °C) and boiling water (100 °C) for 100 times. The devices were immersed to the liquid for 10 seconds each, and kept in ambient air for 10 seconds between immersions. In a final temperature test, the devices were exposed to a constant temperature of 300 °C for 240 hours in dry air.

A commercial shaker (Brüel & Kjaer) was employed to conduct mechanical stress tests investigating vibration-induced failures. The devices underwent vibrations at sweeps between 50 Hz and 2 kHz with a maximum peek acceleration of 200 N/m² (MIL-STD 883 2007 A). An impedance head (Brüel Kjaer Type 8001) induced the vibration and measured the acceleration. The test devices were assembled on a stage fixed to the impedance head. They were tested in horizontal and vertical directions for 72 hours.

3.2 Effect of temperature and vibration on contact resistance

A significant increase of resistance was observed when testing structures were exposed to a temperature of 300 °C. The resistance changed from less than 5 mΩ to more than 70 mΩ averaged levels levels. The elevation of resistance at temperatures 300 °C was observed in five testing devices during a period 12 days (Figure 9).

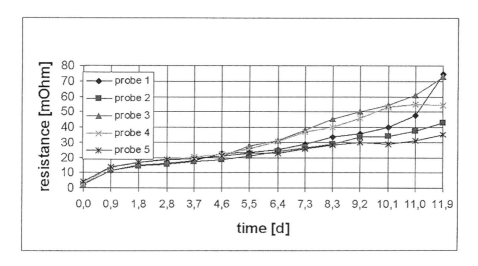

Fig. 9. Absolute changes of MFI interconnect resistance when exposed to 300 °C for twelve days

No major change of resistance was observed during temperature cycling, temperature shock, and exposure to high humidity. The resistance before and after exposure to temperature cycles (n = 45), temperature shocks (n = 18), biased temperature and humidity (n = 45) and mechanical vibration (n = 18) are indicated by bars in Figure 10. Errors bars correspond to standard deviation.

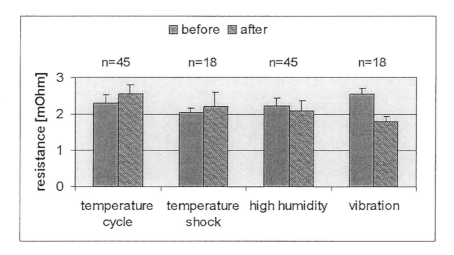

Fig. 10. Mean changes of MFI resistance before and after exposition to test procedures

4. Biomedical applications

4.1 Interconnecting ultrasound transducer arrays

The first system integration has been performed on a demonstrator system for an intelligent cardiac ultrasound catheter. Firstly, the feasibility was studied to connect four dummy ICs - arranged on a squared ceramic carrier - utilizing the MFI technology. A 10 μm thin, μ-flexed cable was designed to wrap around the pad-array site of the chips (Figure 11, left photo). Mechanical and electrical interconnections were achieved at 4 x 24 pad locations. Secondly, four custom designed ICs (1.1 x 2.5 mm² each) were arranged on the same cubic ceramic carrier (Figure 11, right photo). The carrier had a base area of 1.3 x 1.3 mm² and a length of 3.0 mm. 228 interconnects were performed between the pads of the chips. The pads covered an area 95 x 37 μm², each with a center to center pitch of 98 μm. No signs of mechanical stress and electrical interference were observed when the μ-flexed cable was bent at a curvature of 1.0 mm.

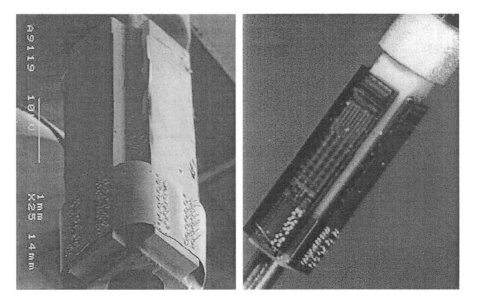

Fig. 11. Micrographs of the MFI technique applied to interconnecting 4 chips on a ceramic carrier. Left: The MFI polyimide cable wraps around 4 dummy chips. Up to 96 sites are interconnected. Right: Tip of an ultrasound cardiac catheter holding 4 IC chips on a ceramic carrier. The MFI flat ribbon embraces the lower part of the chips for interconnecting multiple pads on the IC chips .

In a second application, an interconnection scheme was developed for an dental ultrasound hand-held device for imaging pockets between gums and teeth. A mechanical assembly and electrical interconnection was achieved between an linear array of 110 PZT ultrasound elements and a μ-flexed cable. The PZT elements were arranged at a pitch of 75 μm. The round gold pads on the polyimide cable were 500 nm thick with an outer diameter of 60 μm and $\varnothing$ 35 μm vias (Figure 12). Ball studs were generated using a $\varnothing$ 25 μm gold wire. The studs were riveted through the 10 μm thin polyimide foil connecting to a 2 μm thick gold layer, which had been sputtered and electroplated onto the PZT material.

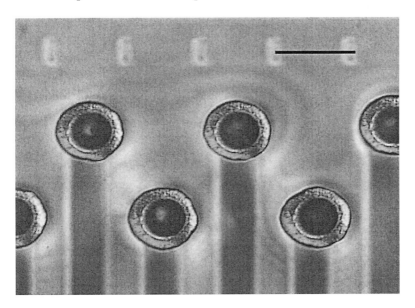

Fig. 12. Micrograph of pads with vias in a polyimide cable. The bar corresponds to 75 μm. The MFI technique was utilized to interconnect 110 PZT elements of an ultrasound array to a PCB.

4.2 Neuroprosthetic Implants

Our group has specialized on the design of flexible neuroprosthetic implant systems which integrate active and passive electrical components on flexible polyimide carriers. The neural devices have been designed for the stimulation of neural cells in the central and peripheral nerve system, in the retina, and in the spine. The MFI technology was applied in all devices, which needed active electrical components close to the stimulating and recording microelectrodes. The fabrication of the devices is based on the previously described photolithographic process. The microfabrication process was extended for applying one technology to embed microelectrodes, conductive tracks, interconnection pads into the polyimide film, to pattern polyimide insulation layers, and to cut and release any shape as carrier from the polyimide film. In the following, two examples of neuroprosthetic implants are given, for which the MFI technology was substantially employed: a retina stimulator device and stimulator for peripheral motor nerves.

4.2.1 Retina Stimulator

The various designs and the functions of epiretinal and subretinal stimulation devices have been previously described [13]. The retina stimulator comprises a receiving unit, a stimulator CMOS based integrated circuit as bare chip, and a microelectrode array that is in direct contact with nerve cells in the retina. The complete packaged system – encapsulated in silicone - is seen in Figure 13. All components have been assembled on a polyimide carrier with patterned gold and platinum layers and a microelectrode configuration to adhere to the retina. The silicone of the receiver unit was molded in the shape of an intraocular lens.

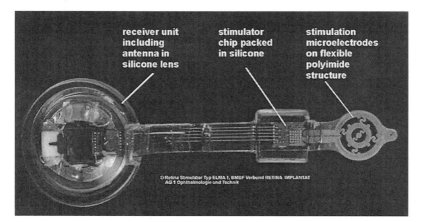

Fig. 13. Retina implant system encapsulated in silicone. The receiver and stimulator chip were interconnected to the flexible polyimide substrate employing the MFI technique.

The receiving unit is composed of a CMOS receiver chip, 1 diode, 1 capacitor, and a receiver coil. The SMD capacitor and SMD diode were soldered onto 300 nm thick Au pads, which had been sputtered on the flexible polyimide substrate. Solder paste (Sn62 Pb36Ag, GLT, Pforzheim Germany) was dispensed on the Au pads. SMD components were aligned and positioned on top of the paste. The PI carrier with the SMD components was placed on a hot plate and heated to 230 °C until the melted solder homogeneously formed around the SMD and on the Au pads.

The bare receiver chip and stimulator chip were mounted and interconnected on the flexible PI substrate employing the MFI technology. The application specific receiving chip decodes the inductively transmitted energy and data signals. The other silicon chip controls the addressing of one of the 24 microelectrodes of the stimulator array. Figure 14 shows the front end of the retina stimulator device with microelectrodes placed on a double ring polyimide structure and the stimulator chip mounted and interconnected using the MFI technology.

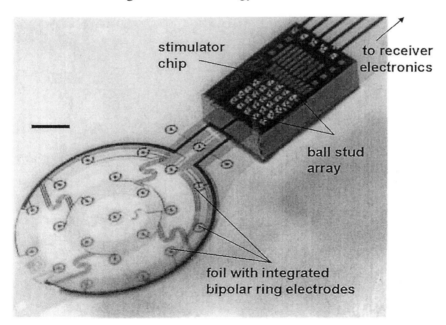

Fig. 14. Micrograph of the retina stimulator tip. The stimulator chip was microflexed to the 20 µm thin polyimide based substrate with embedded stimulation electrodes. The bar corresponds to 1 mm.

The stimulator chip was connected with 25 ball studs (rivets) to the polyimide carrier (Figure 15). Assembling between the connector array of the substrate and a silicon based CMOS integrated circuit has been performed onto platinum pads of 100 μm x 100 μm with a gold wire of 25 μm diameter. The bonding resulted in a gold ball with a diameter of 35 μm. Due to visual inspection during assembling, the yield of the array has been nearly 100% The flexibility of the 20 μm thick polyimide carrier with two embedded and patterned metal layers are demonstrated in Figure 16.

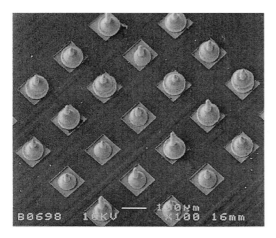

Fig. 15. SEM micrograph of the retina implant ball stud array. Ball studs connect the flexible polyimide carrier with the aluminum pads of the stimulator chip.

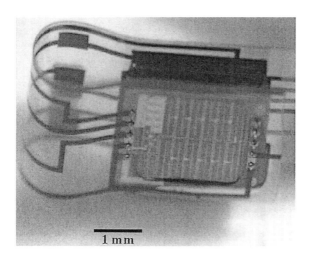

Fig. 16. Bent polyimide substrate holding a microflexed interconnected photocell on one side.

In order to achieve a closer contact of microelectrodes to the tissue, elevated microelectrodes may have an advantage. We have produced protruding microelectrodes by simply stacking the ball studs, which we used in the MFI technology. Four stacked gold studs form a 150 $\mu\mu$m high 'tower' array on a 10 μm thin polyimide foil {Figure 17}. The towers have a pitch of 200 μm. The performance of 'tower' electrodes under in vivo conditions is still awaiting evaluation.

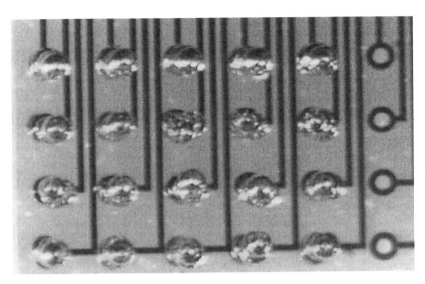

Fig. 17. Polyimide test foil (left) adjacent to the stimulator chip. Test pads interconnect to the aluminum pads on chip for evaluation of electrical signals.

Before the actual fabrication of the final retina stimulator device, test
structure were devised to investigate the reliability of the MFI technology
when in close contact to living tissue. The addition of extra pads on the
substrate allowed testing of electrical functions after assembling and
interconnecting of the bare chips and SMD components. It was found that
an epoxy underfiller was required to avoid moisture and liquid formation
between the chip and the polyimide layer. Capillary force were utilized to
pull epoxy drops through holes in the polyimide foil between the chip and
the substrate (Figure 18).

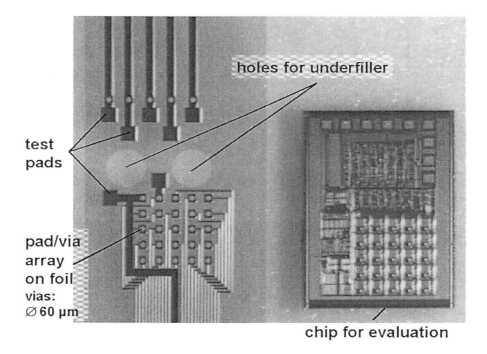

Fig. 18. Stacked gold studs form a 'tower' array on a 10 μm thin polyimide foil.
The pitch between towers is 200 μm. Towers are evaluated for being utilized as
protruding electrodes in a retina stimulator design.

The retina implant system is currently evaluated in animal models. A
passive device – without the electronic components – had stayed for more
than half a year in the eye of a rabbit. The first fully integrated retina
device has been implanted in pig and retina for testing biocompatibility
and electrical function under in vivo conditions.

4.2.2 Peripheral Nerve Stimulator

Cuff-electrodes are commonly used to stimulate peripheral motor nerves or to record from peripheral sensory nerves. The amount of electrodes inside a cuff is conventional limited by the available space for cables and electrodes. We have overcome this limitation by developing an implantable multiplexer system and by integrating 18 to 24 microelectrodes on a polyimide carrier into the silicone cuffs [14]. Only four control lines are needed to electronically control each of the microelectrodes. Figure 19 pictures an MFI assembled and interconnected generic multiplexer system in an unfoldede state. All electronic components of the multiplexer are commercially available. The SMD components were soldered on gold pads embedded in the polyimide carrier. The two multiplexer chips and the binary counter were μ-flexed on the substrate. When folded, the multiplexer was fit into a medical grade silicone tube (25 mm length, 3.5 mm inner diameter, 4 mm outer diameter). For implantation, the tube was filled with two part medical grade silicone (NuSil Med-6015), and was sealed with a silicone adhesive (NuSil Med-1000).

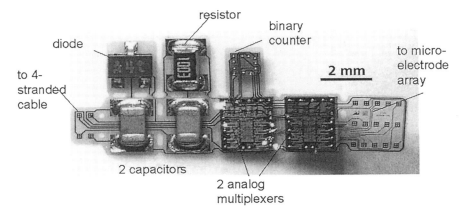

Fig. 19. Peripheral nerve stimulator comprising a cuff electrode array, a multiplexer unit with bare chips and SMD components, and a ceramic substrate interconnecting the multiplexer with a 4-stranded cable

The complete cuff-electrode stimulator system – without tubing encapsulation – is photographed in Figure 20. The system comprises the cuff shaped microelectrode array, the multiplexer unit, and a 4-wire cable for remote control.

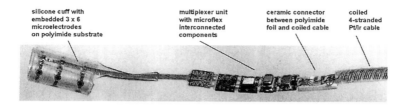

silicone cuff with embedded 3 x 6 microelectrodes on polyimide substrate | multiplexer unit with microflex interconnected components | ceramic connector between polyimide foil and coiled cable | coiled 4-stranded Pt/Ir cable

Fig. 20. Unfolded multiplexer unit exhibiting microflex interconnected multiplexers and SMD components that were solderd onto the polyimide substrate.

For interconnecting the cuff-microelectrodes at one side and the cable at the other side, different interconnection technologies were employed. The 4-stranded cable was welded onto one end of a ceramic carrier. The other end of the ceramic carrier was μ-flexed to the polyimide carrier of the multiplexer system (Figure 21). The polyimide cable extension of the cuff was connected to the polyimide carrier of the multiplexer in a modified MFI technique, employing double-stacked ball studs (Figure 22).

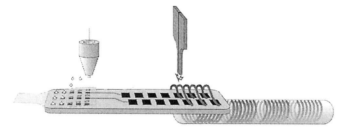

Fig. 21. Schematic of a ceramic substrate on which a polyimide film is 'microflexed' at one side (left) and a 4-stranded cable is welded at the other side (right).

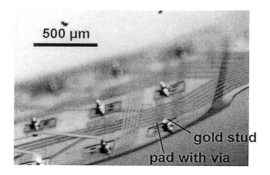

500 µm

gold stud

pad with via

Fig. 22. Interconnection of two polyimide foils by stacked ball studs.

5. Applicability of MFI technology

In need for a new assembling and interconnection technology for a retina implant, we have developed the MicroFlex Interconnection (MFI) technology which enables us to assemble and interconnect bare dice and SMT components on ultra-thin, highly flexible polyimide substrates. Interconnection densities smaller than 100 μm have been achieved. The MFI technique requires minimal interconnection space, corresponding to small space requirements in flip chip technology and chip scale packaging. The MFI technology requires only one bonding process to mount and connect bare chips and other microcomponents to a flexible substrate. Chips and substrates do not need any bumping procedure before assembling and interconnecting. The MFI utilizes simple ball studs from wire bonding as rivets to interconnect the discrete components directly onto the thin polymer substrate. No fiber push connection bonders are needed that go through substrates [4] since our substrates are thinner than 20 μm.

The MIL adopted tests of the MFI technology have demonstrated that MFI is a rather robust and reliable interconnection technology. The increase of resistance during 300 °C temperatures might be explained by the "Kirkendall" effect [15]. The reduced conductivity is most likely due to the diffusion of gold into the aluminum. The Kirkendall effect can be neglected at temperatures below 40 °C. Implants in contact with living tissue always operate at temperatures below 40 °C, therefore no adverse effects due to diffusion of this type are expected. For other applications, the diffusion of gold into the aluminum can possibly be avoided with diffusion barriers in the IC contact pads. Polyimide has been used as a major material for the ultra-thin carrier films and insulation layers. Present trends in micropackaging indicate that copper/benzocyclobutene (Cu/BCB) is used for the low resistivity of Cu and for the well established process to pattern BCB [10]. Copper is a widely used conductor material, also in polyimide substrates [4], [5]. Copper is known to be toxic in contact with living tissue causing severe immunological reactions [16]. We have used platinum and gold as rather inert materials known for their biocompatibility [16]. It should be stated that when copper is appropriately isolated from body fluid, copper is used successfully as implant material, despite its toxic reactivity to body tissue. Donaldson et al. have achieved excellent protection of thick film microelectronic assemblies using one-part silicone adhesives as an encapsulant [17]. We have used non-toxic platinum as conductors since packing of flexible polyimide substrates with silicone rubber has not been tested extensively enough to be sure that body fluids get in conduct with the conductor. The disadvantage of platinum and gold is their higher resistivity in contrast to the one of copper. We are currently looking into electroplating procedures to increase the conductivity in our devices without the use of copper.

Our microelectronic implants on flexible polyimide substrates have been encapsulated in medical grade silicone rubber. Silicone has been reported as suitable encapsulant for implantable microelectronics including solder [18,19]. In order to provide further environmental and dielectric insulation, the use of additional Parylene C coatings will be investigated in future applications. Coatings with parylene have proven to be an excellent choice for brain probes or cochlea ear implants [20].

BCB may prove to be well suited as an alternative material to the polyimide PI 2611. Modified BCB is used as a curing material for certain biomedical implants. Preliminary cytotoxicity tests have shown that the material is non-toxic when used for curing synthetic polyolefins [21]. Extensive tests on the non-toxic effects of BCB are needed. In particular, the tests have to demonstrate non-toxic effects when the material is in contact with highly sensitive neural structures. Furthermore, BCB needs to be proven biocompatible after it has been processed. We learned from our experience with patterning the polyimide 2611, that alterations, for example in gas compositions for reactive ion etching, may lead to a toxic behavior of the patterned material.

Preliminary experiments have shown that polyimide sieve structures with microelectrodces are functional after six month of implantation in animals [22]. The MFI assembled and interconnected retina stimulator is currently tested in animal experiments under operating conditions. It is hoped that the favorable in-vitro and ex-situ results of the retina stimulator device are confirmed in the animal studies. We also apply the MFI technology successfully in other applications, e.g. for cuff-type neural stimulators, for interconnecting miniaturized ultrasound array sensors and for assembly and interconnection of pressure sensors and control chips in active catheters. It should be stated that the presented approaches for interconnections and assembly definitely do not suit for all active biomedical implant applications. The MFI technology is particularly suited for devices that need to be implanted into delicate biological structures with limited space tolerance. The eye is a good example for such stringent requirements.

The present MFI technology relies on a skilled operator to align the structures, and to place the bonds. In order to establish the MFI technology as a packaging and interconnection technology in the semiconductor industry collaborations are needed to automate the procedures and to develop cost-reduced procedures for high-volume production of MFI assembled devices. Flexible circuits made from liquid-crystal polymer (LCP) films have been recently applied to medical devices, as hearing aids and ultrasound transponders [23]. They exhibit low water absorption of less than 0.1 %. Therefore, the material is suited to frequent sterilization. Thin film metallization processes have already been established. Procedures for dry etching LCP films are still under development. Biocompatibility of LCP films needs still to be proven when in close contact with blood, connective tissue, or neural structures.

Conclusions

A new technique for interconnecting microdevices with flexible substrates was developed, termed MicroFlex Interconnection (MFI) technology The MFI technology has been applied to integrate electronics into cardiac catheters, dental ultrasound devices, and prosthetic implants, as nerve stimulators for the retina and for peripheral nerves. The MFI technology allows hybrid integration and assembling of bare silicon dice (ICs) and soldered SMD components at either side of the same flexible substrate. The dice are directly bonded onto the substrates with a 'rivet' that interconnects through the via in the polyimide film onto the chip. The procedure of screen-printing bumps onto the chip or onto the substrate is avoided. Testing according to MIL standards has demonstrated that the MFI technology is robust and reliable. Future work is directed towards the automation of the MFI technology in order to establish a high-density interconnects and packing technology suited for high-volume applications.

We conclude that the MFI technology, comprising highly flexible substrates, high density interconnects, and simple assembly of bare chips and other microcomponents, opens new venues for a novel generation of active biomedical implants with advanced sensing, actuation, and signal processing properties.

174

Bibliography

[1] Thompson, T., Carrasco, A., and Mawer, A. Reliability assessment of a thin (flex) BGA using a polyimide tape substrate. 207-213. 1999. Piscataway, NJ,USA, USA, IEEE. Twenty Fourth IEEE/CPMT International Electronics Manufacturing Technology Symposium / IEEE; Semicond. Equipment and Mater. Int
Twenty Fourth IEEE/CPMT International Electronics Manufacturing Technology Symposium (Cat. No.99CH36330). 18-10-1999.

[2] Arrowsmith, P., Bondi, M., Chen, A., Leblanc, S., Mohabir, R., Panesar, M., Riccitelli, G., and Sterian, I. Early experiences with assembly of chip scale packages. 135-140. 1997. Edina, MN, USA, USA, Surface Mount Technol. Assoc. Proceedings of 4th Annual New and Emerging Technologies for Surface Mounted Electronic Packaging
SMTA National Symposium, Emerging Technologies. Proceeding of the Technical Program. 20-10-1997.

[3] Topper, M., Schaldach, M., Fehlberg, S., Karduck, C., Meinherz, C., Heinricht, K., Bader, V., Hoster, L., Coskina, P., Kloeser, A., Ehrmann, O., and Reichl, H. Chip size package-the option of choice for miniaturized medical devices. 3582, 749-754. 1999. San Diego, CA, USA, USA, SPIE-Int. Soc. Opt. Eng. 1998 International Symposium on Microelectronics / SPIE; IMAPS
Proc. SPIE - Int. Soc. Opt. Eng. (USA). 1-11-1998.

[4] Kasulke, P., Oppert, T., Zakel, E., and Azdasht, G. A low cost manufacturing process for a CSP-flexPAC^{TM}. 83-89. 1997. Berlin, Germany, Germany, Fraunhofer Inst. Reliability & Microintegration. Proceedings of Area Array Packaging Technologies Workshop on Flip Chip and Ball Grid Arrays
Area Array Packaging Technologies. Workshop on Flip Chip, CSP and Ball Grid Arrays. 17-11-1997.

[5] Bergstresser, T. and Sallo, J. S. Flexible laminate substrates for chip scale packaging applications. 1, 143-149. 1998. Northbrook, IL, USA, USA, IPC. Proceedings of IPC Chip Scale/Ball Grid Array National Symposium
IPC Chip Scale and BGA National Symposium. Proceedings. Pursuit of the Perfect Package. 6-5-1999.

[6] Richardson, R. R., Jr., Miller, J. A., and Reichert, W. M. Polyimides as biomaterials: preliminary biocompatibility testing. Biomaterials 14[8], 627-635. 1993.

[7] Mastrototaro, J. J., Massoud, H. Z., Pilkington, T. C., and Ideker, R. E. Rigid and flexible thin-film multielectrode arrays for transmural cardiac recording. IEEE Trans.Biomed.Eng 39[3], 271-279. 1992.

[8]   Haggerty, H. S. and Lusted, H. S. Histological reaction to polyimide films in the cochlea. Acta Otolaryngol. 107[1-2], 13-22. 1989.

[9]   Radlik, W., Zellner, M., Plehnert, K., Heistand, R., Castillo, D., Urscheler, R. Application of photosensitive BCB as interlevel dielectric for a communication MCM-D, Microsystem Technologies '94, Berlin, ISBN 3-8007-2058-2, 1994.

[10] Neirynek, J. M., Gutmann, R. J., and Murarka, S. P. Copper/benzocyclobutene interconnects for sub-100 nm integrated circuit technology: elimination of high-resistivity metallic liners and high-dielectric constant polish stops. Journal of the Electrochemical Society 146[4], 1602-1607. 1999.

[11] Stieglitz, T., Beutel, H., and Meyer, J. U. "Microflex" – A new assembling technique for interconnects. J. Intelligent Material Systems and Structures, 11, 417-425, 2000.

[12] Meyer, J. U., Stieglitz, T., Scholz, O., Haberer, W., and Beutel, H. High density interconnects and flexible hybrid assemblies for active biomedical implants. IEEE Trans. Advanced Packaging, 24, 366-374, 2001.

[13] Meyer, J. U. Retina implant – a BioMEMS challenge. TRANSDUCERS '01 EUROSENSORS XV, the 11th International Conference on Solid-State Sensors and Actuators, Munich, Germany, June 10 – 14, 2001, to be published in Sensors and Actuators, Part A, 2002.

[14] F.J. Rodríguez, D. Ceballos, M. Schuettler, E. Valderrama, T. Stieglitz, and X. Navarro, "Polyimide Cuff Electrodes for Peripheral Nerve Stimulation", J. Neurosci. Met., 98 (2), 105-118, 2000

[15] V. Koeninger, and E. Fromm, "Einflüsse von Al-terung und Kontamination auf die Degradation von Gold-Alu-minium Ballbond", Verbindungs-technik in der Elektronik und Feinwerktechnik; Heft 2, pp. 193-199, 1996

[16] S. S. Stensaas and L. J. Stensaas, "Histopathological evaluation of materials implanted in the cerebral cortex," Acta Neuropath., vol. 41, pp. 145–155, 1978.

[17] P. E. K. Donaldson and B. J.Aylett, "Aspects of silicone rubber as encapsulant for neurological prostheses—Part 2: Adhesion to binary oxides," Med. Biol. Eng. Comput., vol. 33, pp. 285–292, 1995.

[18] P. E. K. Donaldson, "Biomedical engineering—Aspects of silicone rubber as an encapsulant for neurological prostheses—Part 1: Osmosis," Med. Biol. Eng. Comput., vol. 29, pp. 34–39, 1991.

[19] P. E. K. Donaldson "Aspects of silicone rubber as encapsulant for neurological prostheses —Part 3: Adhesion to mixed oxides," Med. Biol. Eng. Comput., vol. 35, pp. 283–286, 1995.

[20] J. Noordegraaf, "Conformal coating using parylene polymers," Med. Device Technol., vol. , 1997.

[21] Fishback, T. L., McMillin, C. R., and Farona, M. F. A new, non-toxic, curing agent for synthetic polyolefins. Biomed.Mater.Eng 2[2], 83-87. 1992.

[22] Navarro, X., Calvet, S., Rodriguez, F. J., Stieglitz, T., Blau, C., Buti, M., Valderrama, E., and Meyer, J. U. Stimulation and recording from regenerated peripheral nerves through polyimide sieve electrodes [In Process Citation]. J.Peripher.Nerv.Syst. 3[2], 91-101. 1998.

[23] Sparrow, N. Electronics open new avenues in device design. European Medical Device Manufacturer. Product Update, 82ff, March/April 2002.

Acknowledgement

Part of the work has been funded by the European Union in the IST program. Funded projects include INTER, GRIP, IMICS, and MEDICS Center of Competence.

Work on retinal implants have been funded by the German Ministry of Research and Education. The author wishes to thank the German retina consortium for their contribution to the project and for the fruitful collaboration.

The author is mostly grateful to his colleagues at the Fraunhofer Institute for Biomedical Engineering (Fraunhofer – IBMT) in St. Ingbert, Germany, for their support, input, and technical ingenuity that accompanied the author's efforts in the area of biomedical microsystems for more than a decade. Among many others, prime persons in the biomedical microtechnology team at IBMT have been Margit BIEHL, Hansjoerg BEUTEL, Thomas DOERGE, Werner HABERER, Joerg HERRMANN, Sascha KAMMER, Klaus-Peter KOCH, Hans RUF, Oliver SCHOLZ, Herbert SCHUCK, Martin SCHUETTLER, and Thomas STIEGLITZ.

Chapter 7

FOLDED FLEX AND OTHER STRUCTURES FOR SYSTEM-IN-A-PACKAGE

Mike Warner and William Carlson
Tessera Technologies

7.0 INTRODUCTION

This chapter describes CSP (Chip Scale Package) variations used in Folded, Stacked, and SIP (System-In-Package) structures. Tape-based CSPs are used for designs requiring folding of the substrate. We will describe examples of production packages, as well as designs existing only as prototypes. Thermal issues are discussed in some detail. Cost considerations are discussed, including breakdown of typical flex tape CSP cost.

This chapter includes a number of design and manufacturing details related to Tessera's μBGA® package family of CSPs, since the authors have extensive engineering and manufacturing experience plus reliability data on these designs. The μBGA® package is currently in volume production by a number of Tessera licensees worldwide. Tessera's most recent packages include a variety of 3-D folded and stacked CSP technologies including SIP customer applications [ref #1]. The conclusions are generally applicable to other CSP package designs, although the details will necessarily differ. A detailed bibliography including URLs to the relevant location within the website is included for those who desire additional information via the convenience and speed of the internet.

7.1 Multi-Die Configurations

The 3-D packaging approach utilizes the vertical dimension, usually with no additional area required on the PCB. Since multiple die are already interconnected, the PCB can usually be simplified and/or reduced in size. This reduction in complexity facilitates the move towards 2-layer boards (many today are 8 or more layers), further lowering overall cost of the final assembly.

7.1.1 Chip Scale Packages

CSPs (Chip Scale Packages) are the preferred building blocks for packaging of multi-chip devices. The usual rule of thumb is that packaged area of a CSP is no more than 120% of the die itself. Examples include a number of proprietary designs including Tessera's μBGA® package. Intel adopted this package in 1998 for FLASH memory as it reduced footprint by 80% compared to the equivalent capacity TSOP [ref #2]. One CSP benefit in general (and μBGA® package in particular) is improved electrical performance due to minimized lead lengths. Although this was not a critical issue for initial CSP implementation for memory die, this electrical advantage soon became an enabling technology for subsequent higher performance applications, such as the Rambus RDRAM® modules used in Pentium-4 motherboards.

7.1.2 Folded and Stacked Package Concepts

Tessera has developed a number of 3-D packages based on variations of the μBGA® package including Folded, Stacked, and various hybrids. The primary 3-D advantage is a significant reduction of footprint and volume over single-die implementations, since a multi-die package consumes about the same PCB real estate as a single-die CSP. The height is typically equivalent to single die packages due to use of thinner die. This highly space-efficient configuration is very useful for applications with size and weight limitations such as cellular phones, PDA, and thin laptops. An unexpected but logical application includes Internet "server farms" where very high facilities cost per square foot help justify the most compact hardware designs. The stacked package allows equipment designers to increase capacity and/or functionality in the same volume, often facilitating new capabilities (e.g. 3rd generation PDA with telephone, internet, GSM, and photographic capability). Intel was again a leader in this field, an early adopter of a multiple-die folded design, which afforded a 4X improvement over a single CSP, or 20X compared to the original TSOP [ref #2].

7.1.3 Evolution to System in a package

SOC or "Systems-on-Chip" employs technologically compatible functions integrated into a single die. When different technologies cannot be included on a single die (e.g. mix of silicon and gallium arsenide, digital plus analog, high voltage and small signal). The SIP or "System In Package" allows a mix of functionally related but differently processed die into an integrated sub-system. The merged parts can also include passives and I/O functions (e.g. antenna, microphone). Additional SIP advantages include the ability to use established die designs for faster time to market, plus each sub-assembly can be independently tested. The SIP devices can then be stacked in combination with other devices (e.g. memory, ASIC). The end product has most die interconnects within the SIP assembly, so the number of surface mount connections on the final PCB are minimized.

The photo above shows stacked CSP packages mounted on a common flex interposer with solder mask on both sides. In this implementation a microprocessor is mounted on the bottom of the flex (not visible), a DRAM, SRAM and an EEPROM are mounted on the side shown. The bottom has 106 I/O balls surrounding the microprocessor and nearly 400 interconnects are made in the 0.9 MM thick System-In-Package. A layer of solder mask is added to both sides of the flex interposer.

MP3 player SIP illustrated below combines memory plus encoder and decoder in one package.

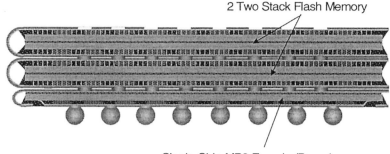

2 Two Stack Flash Memory

Single Chip MP3 Encoder/Decoder

7.2 CSP variations as Building Blocks

There are numerous ways to both build CSPs and to combine them into integrated packages. This section will cover several currently used and proposed multi-chip designs. Included are SIPs with CSPs mounted "side by side", "folded", "stacked", or combinations of techniques. Use of thin die (currently in volume production down to 114 microns) has allowed the SIP to maintain the thickness of a conventional single die package, usually less than 1.2 mm in thickness. Die cannot be made arbitrarily thin, however, since localized stresses on the circuit side of the die due to sandwiches of different materials can cause die warping if wafer stiffness is insufficient to resist these stresses.

7.2.1 CSPs on Substrates

The addition of a redistribution layer between die and PCB solves a number of electrical issues at some added cost. Redistribution fixes the pad-pitch discrepancy between die and PCB by bonding die pads to an intermediate redistribution layer, which in turn provides electrical connections to a ball-grid array for PCB mounting. This layer typically provides considerable latitude in routing die I/O to accommodate circuit design issues (e.g. isolating power, ground, inputs, and outputs). Some die layouts may be incompatible with PCB layout using single layer redistribution if leads must cross over each other. Use of a 2-metal substrate (e.g. similar to a 2-sided PCB) allows crossovers, which solves this problem, but at additional cost.

Redistribution of electrical contacts allows a functionally identical but physically different die to share a common device level application. This facilitates 2nd sources and avoids need for die or PCB redesign, although a custom redistribution layer is typically required for each CSP.

Flexible substrates are typically based on Polyimide films for mechanical strength and temperature resistance. The basic structure is nitrogen containing organic polymer. The most commonly used versions in semiconductor packaging are those similar to Dupont's "Kapton" [ref #3], or Ube's "Upilex" [ref #4]. Superior physical properties come from strong charge-transfer bonding between the nitrogen and carbonyl group, which is also effective between adjacent strands of polymers in the bulk material. This 3-dimensional bonding serves to hold the material together in all directions with great strength [ref #5].

polyimides may stack like this

Polyimide provides excellent electrical insulation, chemical stability, supports fine lines and spaces, has temperature stability up to 400°C, provides alpha ray shielding, and has excellent flexibility in thin films. Single metal layer polyimide tape typically sells for about 8 cents per square centimeter in metalized TAB tape form.

182

7.2.2 MicroBGA®

Tessera solved the CTE mismatch problem between silicon and the PCB with a structure utilizing TAB tape as a redistribution substrate plus a silicone "compliant layer" between the die and tape to mechanically decouple them. This flexible attachment improves reliability by allowing relative movement caused by thermal expansion mismatch between die (CTE of 3 ppm) and PCB (FR-4 at 18 ppm). Dow Corning and others developed flexible silicones for packaging, including adhesives, encapsulants, and spacers [ref #6]. Electrical connections between die and substrate include an S-shaped flexible lead that is allowed to move within the compliant layer. A quantitative example of the CTE issue is a 10mm wide die cooling from solder reflow "freezing point" of 183 degrees centigrade to room temperature. The relative movement is 24 micron (10mm * (18-3ppm) *10^-6*(183-25)), which leads to substantial stress at the solder joint. The situation is aggravated with thermal cycling, where repetitive stress application causes fatigue at the solder joint, leading to cracks and breaks. The compliant layer solution typically shows service for 2,000-10,000 thermal cycles for virtually all die sizes.

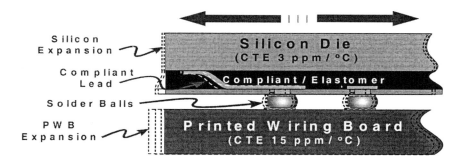

The standard µBGA® package utilizes single metal polyimide (TAB) tape, with "flying leads" across a window in the substrate. The construction can be either "circuits-in" where the metalized surface faces the die, or "circuits-out" where the circuitry metal faces outward. The "circuits-in" approach is generally favored as a lower cost approach, having fewer manufacturing steps since the polyimide tape also serves as a solder mask. [ref #6]

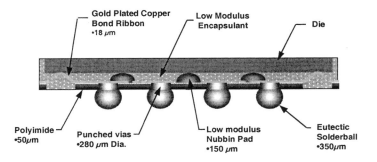

Gold Plated Copper Bond Ribbon •18 µm

Low Modulus Encapsulant

Die

Polyimide •50µm

Punched vias •280 µm Dia.

Low modulus Nubbin Pad •150 µm

Eutectic Solderball •350µm

One end of the flying leads contains a notch that allows controlled breaking of the lead by the thermosonic bond tool during the bonding process. Also during this process the lead is configured into an S-shape by the bonder. This scheme provides the minimum lead length between die and substrate, typically under 0.4mm. The structure becomes challenging when die pads are on all sides including the corners, as there is no way to support the package once the leads are bonded. Crossovers and very dense routing may require two metal tape. Tessera has built CSPs with up to 768 I/O on a 0.5 mm area array pitch.

A typical µBGA® package DRAM structure:

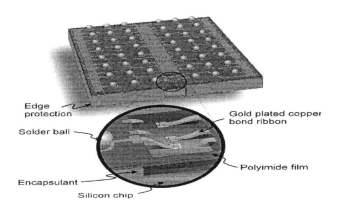

Edge protection

Solder ball

Encapsulant

Silicon chip

Gold plated copper bond ribbon

Polyimide film

7.2.3 LGA - Land Grid Array

A variation of the standard μBGA® package is the "Land Grid Array" which employs Solder Lands instead of solder balls for package-to-package interconnects. The amount of solder is reduced to keep the solder within the bounds of the vias, which reduces the thickness of the package.

The low stress compliant layer allows thinner die to be packaged in a very low profile package (less than 0.3mm) for use in SIP applications.

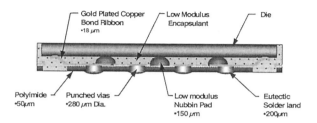

7.2.5 FμBGA® - Wire Bonded

This CSP is a wire-bonded variation of the μBGA® package, offered to the industry through cooperation between Tessera and Toshiba. The die is face down, electrical connections face the tape and die and are bonded through a window in the tape required for bonding tool access. Wire length in this case is approximately 0.4mm in length. The following two views illustrate construction.

Gold wire is an additional cost in this structure. The use of wire bonding is traditional in the packaging industry, making this version attractive to

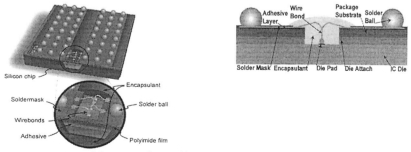

many assemblers. Overall cost difference is usually insignificant between lead-bond and wire-bond methods.

7.2.6 Flip Chip

Flip chip has a direct solder bond between the Silicon die and the redistribution layer. During thermal cycling, the difference in expansion rates results in unrelieved mechanical stress and an increase in failure rate. A typical failure point is between 600-800 cycles using a 125 °C temperature change depending on die size. This is a significant concern when a wide range of temperature exposure (e.g. -40 to 125 °C.) is anticipated, with large die, or for mission-critical applications. Flip chip applications typically underfill the die to help distribute the stresses. Underfill is an extra expense during manufacturing. Nevertheless, CSP packages containing flip chip die are used in SIP applications. (See section 7.7.1.3.)

The two illustrations below show the stresses due to thermal cycling on solderballs with and without stress-relieving layer.

(at -65°C end of "883" Condition C cycle)

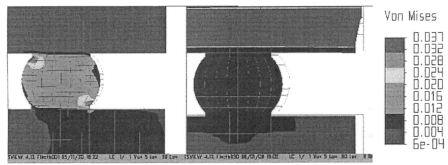

Flipped Chip w/o Underfill   Compliant μBGA CSP

7.3 Mechanical Construction

CSPs are ideal building blocks since they take up minimum space and therefore represent the most space-efficient components beyond bare die. Due to robust materials of construction and encapsulation of the die, CSP devices are mechanically rugged, and can be made very resistant to hostile environments. The μBGA® package, for example, fully meets JEDEC Level-1 requirements. Due to short lead lengths, inductance and capacitance is minimized, allowing additional values to accumulate from the multi-die packaging (e.g. leads along tape) without exceeding maximums specified for the entire assembly.

Small size itself can be an enabling technology where absolute minimum volume is required, or where additional complexity must fit within the same available space. Examples include implanted medical devices such as heart pacemakers/stimulators and cochlear hearing restoration, plus the ever-shrinking form-factor PDA and cellular phone. The small-scale attributes of CSP devices also provide electrical advantages to help enable high frequency applications. A case in point is Rambus memory modules, where the clock rate of 800MHz and critical timing could not be supported by TSOP or QFP structures due to added inductance and capacitance contributions from the lead-frames in these conventional packages.

Tape based structures have the unique advantage of being able to be twisted and/or folded, while at the same time performing an interconnect function, both from die to die, and to contacts intended for an external circuit (e.g. Ball Grid Array to PCB). As in origami, variations are limited only by one's imagination and being able to build it. Herewith are some of the more interesting designs and construction details.

7.3.1 Single Fold, Single Layer

This design is a "building block" for stacked devices using folded tape to create interconnects on both sides of the subassembly. This allows multiple numbers of the devices to be stacked, with a PCB connection on the bottom part. Three such subassemblies include a simple 1-fold wrap-around, a 2-fold wrap-around (advantage of shorter lead lengths), and a version with multiple die.

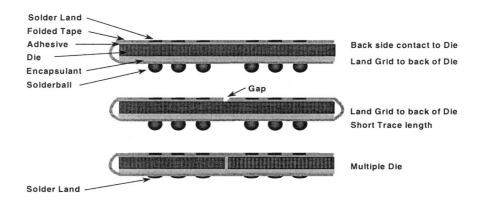

7.3.2 Superimposed Die on Folded Tape

The in-line folded die design places die side by side along a strip of tape, then folding one over another uses a back-to-back die adhesive attachment. In the case of a 4-die model, die at each end are folded over their neighboring central die. Then these two-die stacks are folded again to form a single vertical stack of four. As in other technologies, the "Devil is in the details". The devilish details include:

- Thin Die (typical 114μ or less) for low overall Z-height (typical <1.2mm)
- Thin tape (25μm) since it enters into package height multiple times
- Die-to-Die adhesive thickness (12μ), material, assembly processes
- Shape of folded section (service loop), avoid kinks and sharp bends
- Encapsulation, filled vs non-filled folded "service loop" section
- Heat transfer within the stack, and to external PCB
- Signal cross-talk between various elements of the stacked package
- Circuit path length considerations between ends of tape
- Stack rework impractical after die to die bonding
- Potential multiple reflow operations during build-up of the stack
- Yield challenges, when to use "Known Good Die"
- Cost tradeoffs, complex package vs simplicity of final installation

The starting point is TAB tape with interconnect circuitry provided by the tape supplier. Typically the metal layer is on one side only, with vias to the opposite side of the central die for a ball grid array.

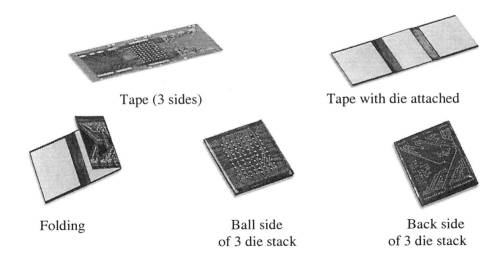

Tape (3 sides)   Tape with die attached

Folding   Ball side
of 3 die stack   Back side
of 3 die stack

188

The tape layout must include sufficient slack for the folding operations, including extra distance for the final central fold, which must span the edges of all four die. Die are adhesively bonded to the tape sites, then electrically connected (lead bonded or wire bonded), and finally underfilled or over molded to protect the leads from damage during subsequent processing.

The folding process is done robotically in a fixture providing approximately a 12-25 micron adhesive thickness between adjacent layers. After adhesive application, the folding tool aligns the two components to be joined, and applies sufficient heat to achieve a first stage cure, after which the part can be handled without risk of separation. Any secondary folding (e.g. 3 or more die) is accomplished in the same fashion, after which the part can be processed (e.g. solder balls applied, inspection) for final use. The finished die stack can now be tested and/or mounted to another sub-assembly, or to the finished product PCB.

Cut-away views of typical versions: Two Die Single Fold and Three Die Double Fold

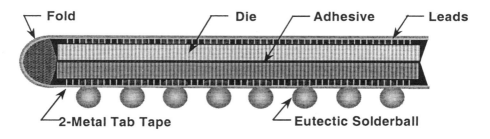

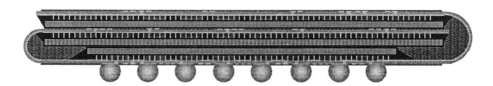

One unique technical issue for folded devices is contour control of the tape fold itself, which helps control stresses associated with thermal expansion. If the bend area is filled with encapsulant having a high thermal expansion coefficient (such as silicone) the bend area will "balloon" during thermal cycling, putting stress on the tape and its circuits. Copper circuitry can work harden, so continued flexing due to thermal cycling can lead to stress cracks and eventual intermittent electrical failures unless package is properly designed.

7.3.3 Superimposed Subassemblies

The folded or stacked devices can be combined into "stacks of stacks" for greater functionality in the same footprint. There are a number of options including flexible and rigid substrates incorporated within the stack.

"Folded and Stacked" examples below includes stacked assemblies of 2 or more folded packages. One advantage is the ability to pre-test the subassemblies before final assembly ("Known Good Die" analog). Another advantage is to inventory only one subassembly (the 2 die folded), from which any number can be assembled into a customized final configuration (e.g. 2, 4, 6, 8 or more die packaged together). This technique is attractive for memory expansion, where capacity can be increased on the same PCB by simply increasing the number of folded die subassemblies per stack.

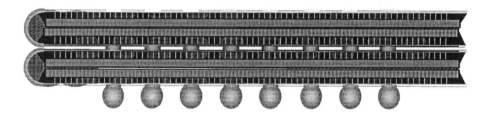

Double 2-Die Folded Package Construction
0.7mm overall thickness (2-die)
1.2mm overall thickness (4-die)
1.7mm overall thickness (6-die)

7.3.4 Ball Stacked Devices

"Ball Stacked" structures involve die mounted on a rigid or flexible substrate which are then combined using solder ball connections around the perimeter of the substrate. Die to substrate connections are made via gold wire bonding or lead bonding (μBGA® package version shown), followed by underfill or overmolding each package. Solderballs are attached and reflowed to each subassembly. The individual die subassemblies can then be vertically aligned and reflowed again to provide both electrical and mechanical connections between them. The bottom member of the stack may also have an array of external solder balls for final connection to the application PCB. This is particularly useful in meeting JEDIC standard footprints or where additional heat transfer is required.

A Center-Bonded version is also available, termed "μZ ®" stacked package. The building block of this stack is a μBGA® package, with attachment balls at the perimeter of the substrate.

Any number of these building blocks can be stacked in a vertical array to provide denser packaging. One typical example is memory die, where attachment balls can be selectively omitted to distinguish one die from another for memory addressing in the finished multi-die module. In the typical case pre-attached solder balls are on each subassembly, then the stacked array is subjected to a secondary reflow in a holding fixture which makes the interconnects.

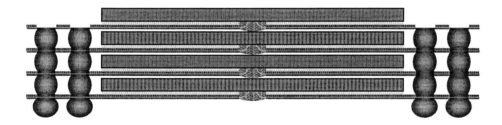

Packages requiring multiple reflow operations have small risk of interconnect defects with properly controlled reflow processes. Small

amounts of relative movement between components of the stack are possible when solder is liquid. Fortunately surface tension forces tend to keep pads on either side of a solder joint aligned and the stack pulled together, however, unintended mechanical forces have the potential to misalign members of the stack during reflow.

7.3.5 Hybrid and Unconventional Structures

With a number of different options in packaging and assembly methods, it's inevitable that combinations of technologies will emerge where they provide unique advantages not available with a single method. A few examples and some detail on the more promising ones:
- Combination of rigid and flex substrate subassemblies
- Combination of Wire-Bond and Lead-Bond structures
- Combinations of die connected via solder plus other methods
- Stacks of Stacks (folded or stacked subassemblies bonded together)
- "Possum", a small die directly attached to a larger one
- Flap & Stack, providing multiple contact areas
- Geometric pattern folds (square, pentagon, hexagon, octagon, etc.)
- Wrap-Around "candy wrapper" or "Jelly Roll" designs

7.3.5.1 "Combo"

The combo stack utilizes two different assembly technologies; a face down center bonded die attached to the substrate via lead bonding, with a second die attached back-to-back and wire bonded to the same substrate. This technique allows two same-size die to be separated only by a bond line for minimum stack height. The top die usually has bond pads at the edges. The entire assembly is then encapsulated via over-molding. This stacked package design has advantages when combining center bond pad die and peripheral bond pad die in a single package.

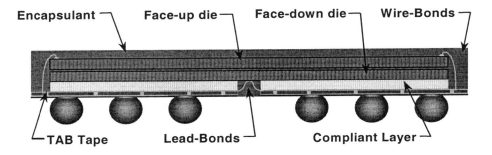

7.3.5.2 "Possum"

The "Possum" stack utilizes a small thin die attached to the underside of a larger one, (think of a mother Opossum carrying her young) then the pair attached to the PCB via solder balls on the larger die. Attachment between the two die is typically via reflow. This design takes ZERO volume beyond that required of the larger die, but requires a thin lower die and minimum height solder attachment, since all can be made to fit within the package ball height dimension (after reflow) of approximately 675 microns.

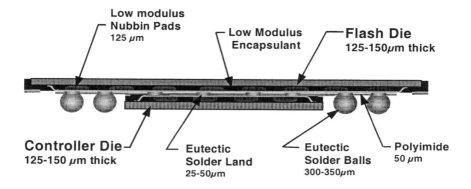

7.3.5.3 "Flap & Stack"

"Flap and Stack" or "Folded LGA" consists of two pairs of die folded together, then reflow-attached to a larger die via a "flap" of polyimide which has been folded over the larger die for interconnect purposes. A practical example includes two SRAM die and two Flash die placed over an RF Baseband die for an integrated radio subsystem. An example of this structure was presented at a recent IMAPS conference [ref #7], where the following "case study" was discussed.

Case Study- The Tessera design engineering team, working closely with an independent substrate supplier and the company's own assembly process development team, has developed a number of multiple-die System-In-Package concepts. A five-die package recently developed for a customer by Tessera furnished a single ASIC and four memory functions in a single low profile fine-pitch array package outline. The ASIC substrate profile follows the die outline on three of the four sides, extending beyond the die on one side to furnish an array interface to mate with the memory assembly. The contact array that will eventually mate with the circuit board is on the opposite surface of the die.

To accomplish this merging of memory and ASIC functions, the assembly is processed by two separate operations. In the first stage, the memory die are assembled (face-down) onto one flexible substrate and electrically tested through a contact matrix on the outer surface. The memory package is then folded in half onto itself (two die on each half) and secured with a compound material. In a separate operation, the ASIC is assembled to its own flexible substrate.

Following the testing of the ASIC assembly, the flexible substrate extension folds onto the top surface of the die and is secured with the bonding compound. The two assemblies are aligned and joined using a reflow solder process. The finished five-die package (illustrated in Figure 1) can then be transferred into final test and with solder balls applied to the bottom contact array, made ready for PCB assembly.

Baseband ASIC + 2-Flash + 2-SRAM
Folded/LGA Package Construction
1.2mm overall thickness (5-die stack)

Shown is the five-die folded-flex stacked µZ-F™ package outline. This package has a low 1.2mm in overall height and is only slightly larger than the largest die outline.

The folded-flex stacked-die package is just one of many innovations developed by Tessera engineers. The most significant result achieved in the development of µZ-Folded and stacked package technology is the dramatically small outline, a real benefit to developers of miniature hand-held electronic products. The package technology not only offers the smallest overall outline possible, the unique combination of materials, thin die, lead or wire-bonding and folding methodology furnishes one of the lowest profile multiple die package systems available.

7.3.5.4 Multi-Folded

Packages with multiple folds multiply space saving benefits by including a large number of die on a single substrate. So many fold geometries can be utilized that the term Origami Packaging is often used to describe this technique.

7.4 Stack-Up Height

The overall height of a stacked or folded package is typically 1.0 to 1.5 mm, and depends significantly on those items which contribute more than once to the overall height (e.g. die and tape). A typical sub-assembly layer in a stack follows:

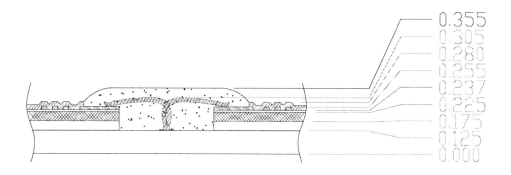

Required solderball height for stacking is 250μm

If, for example, 3 of these subassemblies were stacked into a single module the post-reflow package thickness would be:

3 Subassemblies at 0.355 mm	= 1.065 mm
2 bondlines between die at 0.025 mm	= 0.050 mm
Solder Ball Attachment	= 0.250 mm
Total Package (after reflow)	**= 1.365 mm**

The solder ball interconnects are a significant part (about 20%) of the overall package height, and need to be accounted for accurately. Solder ball diameters typically used are between 300 and 500 micron.

The spherical ball wets the solder pad during reflow and becomes a shorter hemisphere. The solder attachment width conforms to the wetted surface of the solder pad, and the balance of solder-ball volume defines the post-reflow height, at about 80% of the original ball diameter for a 400 micron solder ball.

The solder hemisphere becomes a barrel-shaped column of solder between devices (or device and PCB) after attachment via reflow. For a typical solder ball of 400 micron and pads of about 320 micron, the solder maximum column width is about 400 micron, and height is calculated to be about 260 micron.

Self-Alignment is a valuable attribute of solder reflow for interconnects or mounting devices to a PCB. "Surface Tension" refers to the natural forces minimizing surface area. The molecules at the surface of a liquid are missing outward facing bonds, which is a higher energy state. The lowest energy (most stable) configuration has therefore minimum surface area, which generally favors spherical surfaces. In the case of a solder column, the minimum area occurs when the two substrates are vertically aligned. When the solder pads are offset, the solder column is "tilted" and surface tension forces from all interconnects combine to pull the substrates into closer alignment. These forces are surprisingly strong. Blaupunkt prepared a video showing that surface tension forces during solder reflow of a BGA are sufficient to pull devices up an inclined plane.

Self-clamping is another benefit related to surface tension. When the packaging components mechanically interfere before the solder column is minimized in height, surface tension applies a net force pulling things closer together. The effect is a natural and gentle compression of packaging height during reflow. Upon cooling, the solder contracts less than the remaining package, material dependant, resulting in a slight gap between stacked packages. The combined natural benefits of compression and alignment favor well-aligned and compact stacked structures

7.4.1 Assembly Height Variations

Lead bonded versions of stacked or folded die have an intrinsically lower height than wire bonded versions, since lead bond connections from die to tape are contained entirely within the space between them.

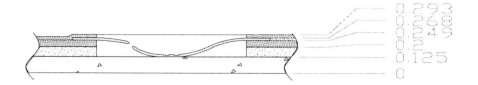

For the wire bonded version, additional space must be allocated for the bond wires rising over the top surface of the die, plus overmolding required to protect the gold wire leads from damage during subsequent processing. A folded package will typically contain 2 to 4 die, so the difference between these two techniques must be multiplied accordingly, accounting for 0.25 to 0.5 millimeter additional height using the wire bonded approach.

7.4.1.1 Thin Die Considerations

A significant component of SIP overall height is the repeated contribution of thickness from multiple stacked or folded parts (e.g. tape, die, adhesives). Thus in a 4-stack, the effect of die height is multiplied by 4, making it a prime target for reduced thickness. The current state of art for volume production wafer thickness is about 114 microns (4.5 mils). The good news about thin die is increased flexibility. Heat generated by operation of the die circuitry can flow more quickly through a thinner die cross section, improving cooling possibilities. Bad news includes more stringent polishing requirements, since chips and cracks (or other points of stress) represent a larger proportion of the thin cross-section and therefore must be closely controlled to avoid crack propagation and broken die. If the die is "too thin", stresses from dissimilar materials on the die face circuitry may mechanically distort the die, resulting in loss of planarity and possible manufacturing difficulties.

7.4.1.2 TAB tape thickness options

Common polyimide tape substrate thickness includes 25, 50, and 75 microns. The thicker formats are stiffer and therefore easier to process, while the thinner versions reduce contributions to overall package height and are easier to fold. Empirical data shows the 25 micron tape thickness requires only 31% of the force required to bend 50 micron tape due to decreased stiffness. Many folded packages therefore employ a 25 micron tape thickness. The metal layer is typically 18 microns of soft copper with a 0.5 to 2 micron gold overcoat. The metal layer adds significant stiffness to the polyimide tape, making the thinner substrate even more desirable.

7.4.1.3 Die to Die Adhesives

A small bond line of adhesive is required to keep the folds in place. The adhesive is typically dispensed on the back side of one die and the back side of a second die placed in contact with that of the first during a folding operation. The subassembly is heated for a first-stage cure to hold the parts in place, in preparation for additional folding and a final cure. The bond line thickness is typically 12-35 microns, Tessera has used both epoxy and silicone based adhesives with excellent results.

7.5 ELECTRICAL

Tessera and others have done extensive electromagnetic and circuitry simulations for detailed assessments of effective capacitance and inductive values due to construction techniques. The level of detail involved would be inappropriate for a packaging overview, so this section's focus will be on a few items of particular relevance to small scale assemblies.

7.5.1 Lead Lengths

Short lead length provides a significant performance advantage for Chip Scale devices by minimizing distance between the substrate and die electrical pads. With a folded design incorporating several devices, some lead lengths may increase to the maximum length of the folded tape length but normally significantly less than individual packages on a PCB. This is unimportant for relatively low speed devices (e.g. FLASH memory), but may be a significant issue for high speed devices sensitive to inductance or propagation delay, and points out the importance of the physical arrangement of die within the folded structure.

7.5.2 Passive Devices

Passive elements are considered part of any electrical circuit design but will have effective values modified by the packaging since the leads and layers of devices contribute small amounts of inductance, capacitance and electrical resistance. At high frequencies and low voltages common in modern packages, precise control of passive values requires including the package contribution. In some cases the overall package design itself can include implementation of passives either as discretes or integrated on a silicon or glass substrate.

7.5.3 Computer Simulation

Computer modeling is the primary method of package designing for the small-scale environment. Numerous software tools are available and material properties are well understood. Tessera follows the typical iterative practice of modeling, prototyping and measuring , then putting the results back into the model for closer refinements. Typical material properties include:

Element	Material	Electrical Property	References
Die	Silicon	$\varepsilon_r = 11.9$, $\mu_r = 1.0$, $\tan\delta = 0$	Maxwell® SI S/W Internal Database
Trace	Copper	$\sigma = 5.8e+7$ S/m	Maxwell® SI S/W Internal Database
Lead	Copper	$\sigma = 5.8e+7$ S/m	Maxwell® SI S/W Internal Database
Solder ball	Solder	$\sigma = 7.0e+6$ S/m	Maxwell® SI S/W Internal Database
Solder mask	Taiyo PSR4000BV	$\varepsilon_r = 4.71$, $\mu_r = 1.0$, $\tan\delta = 0.0332$ @1 MHz	*Taiyo PSR-4000BV (AUSS) Green Liquid Photoimageable Solder Mask for BGA Application: Data Sheet*, Dec. 1999.
Polyimide tape	Dupont Kapton film	$\varepsilon_r = 3.3$, $\mu_r = 1.0$, $\tan\delta = 0.006$ @100 KHz	*SRC/CIDAS Database*, Jan. 2000.
Elastomer	Dow Corning 6811	$\varepsilon_r = 2.9$, $\mu_r = 1.0$, $\tan\delta = 0.0002$ @1 MHz	*Dow Corning Encapsulant Products Data Sheet*, 2000.
PCB dielectric	FR4	$\varepsilon_r = 4.3$, $\mu_r = 1.0$, $\tan\delta = 0.04$ @1 GHz	EIA/JEDEC Publication, *EIA/JEP126*, May 1996. *SRC/CIDAS Database*, Jan. 2000.
PCB ground plane	Copper	$\sigma = 5.7e+7$ S/m	EIA/JEDEC Publication, *EIA/JEP126*, May 1996.

7.5.4 Electrical Interconnects

One of the most powerful advantages of SIP designs is the ability to interconnect the die internally to the package and thus reduce the number of package ballouts. In many SIP applications 400-600 internal interconnects are made. This results in increased performance; simpler PCBs usually lower cost and always much smaller overall size.

7.5.4 Alpha Radiation

Intel and IBM observed electrical reliability problems [ref 8, 9] traced to sources of ionizing radiation. Trace amounts of radioactive contamination are part of lead used for solder alloys and "natural" materials used in packaging. Radioactive decay of Uranium and Thorium involves a series of transformations ending at inert lead isotope Pb-206. Uranium-238 has a half-life of 4.5 billion years and Thorium-232 is 14.5 billion. This compares to 4.5 billion years for the age of the earth, so these sources of radioactive lead are not going away anytime soon. Along the decay path of both materials, radioactive lead Pb-210 is constantly being formed with a 21-year half-life. Lead's disintegration sequence includes alpha particles (helium nuclei) with an energy content of about 4 MEV (Million Electron Volts) compared to 3.6 EV (Electron Volts) needed to generate an electron-hole pair in silicon. One such alpha particle can therefore generate millions of carriers (**ref #10**), potentially exceeding charge content of RAM memory cells, thereby leading to transient memory errors.

Chemical refining cannot distinguish between isotopes of lead. When natural lead is recovered from ore, it always contains a small amount of the radioactive isotope. Once lead is isolated from sources of Pb-210, one can simply wait until the existing Pb-210 content sufficiently decays to relative insignificance. One technique method utilizes "Old Lead" from 300 year old cannon balls or lead artifacts, since radioactivity of Pb-210 in natural lead will have decreased by a factor of 20,000. (14.3 half-lives of 21 years). Isotope separation is commercially done [**ref #11**], at some significant cost premium. Fortunately there are no naturally occurring radioactive Tin isotopes, so lead presents a singular problem for Sn-Pb solder.

Radioactive contamination is discussed in J-STD-012 (**ref #10**) where Flip-Chip is noted to be particularly vulnerable due to direct solder contact with the die. According to JEDEC, "Circuits sensitive to soft errors are DRAM, closely followed by SRAMs, and some flip-flop circuitry". "Sensitivity increases as chip geometries shrink and device (node) critical charge diminishes." Fixes include use of thin shields (polyimide is typical), or elimination of radioactive contaminants. A chart of "stopping power" shows that even a small amount of any solid material is an effective alpha stopper [**ref #12**].

AMOUNT OF MATERIAL REQUIRED TO STOP ALPHA PARTICLES

Radioactive Element	Atomic Number	Atomic Mass	Radioactive Half Life	Energy MeV	<--Microns required to stop Alpha Particles-->			
					Air (80%N_2)	Mica	Aluminum	Gold
Thorium	90	232	1.4x10^10yr	4.0	28000	14	16	6
Radium	88	226	1620 year	4.8	33000	17	19	7
Thorium	90	228	1.9 year	5.4	39000	20	23	8
Polonium	84	218	3.0 minutes	6.0	46000	23	27	9
Polonium	84	216	0.16 seconds	6.9	56000	28	33	11
Polonium	84	214	1.6x10-4 sec.	7.7	69000	35	41	14
Polonium	84	212	3.0x10-7 sec.	8.8	86000	43	51	18

Common Mica = Muscovite mineral = $H_2KAl_3(SiO_4)_3$, density 2.8, used for RF capacitors

The problem is not limited to lead since all "natural" materials have small amounts of radioactive contamination. Dow Corning adopted synthetic fillers for CSP silicones since natural materials cannot be sufficiently decontaminated to low levels of α-emission required. The Tessera μBGA® package is particularly immune, using the Dow Corning encapsulant, plus polyimide tape as a "shield" from potential solder-based radiation.[ref 14].

Some literature values as alpha emissions per cm^2 per hour:

α-emissions per square-cm per hour	Min	Max	Typical
High Lead Solder	0.05	10	5
Natural Aluminum Oxide			0.1
Plastic Materials			0.04
Silicon Wafer			0.004

7.6 Thermal Considerations

Thermal issues are becoming a greater overall design consideration, since frequencies and heat dissipation of devices are going up. ITRS [ref 13] forecasts microprocessor power consumption in high-performance systems going from 140 watts in 2002 (at 2.3 GHz) to 288 watts in 2016 (at 6.7 GHz). During this same period core voltages will drop from about 1.0 volt to 0.4 volts, but the maximum junction temperature of Silicon must remain about the same at 85 degrees centigrade. This implies enormous current (0.4 volts requires 720 amps to achieve 288 watts), and power dissipation scales by the square of current (I^2R). At the same time other computer functions (memory, video, communications, wireless) will become bigger and/or faster, increasing the overall quantity of electronics, adding to the heat load. As device packaging becomes increasingly dense, heat management will grow in importance. Active die (plus passives) sandwiched between electrical insulating materials in a folded or stacked package can develop "hot spots," since electrical insulators tend to be thermal insulators as well. Several techniques are available to redistribute heat and facilitate its removal. Convection cooling with fans (which contribute their own heat plus noise) has significant heat removal limitations which eventually must yield to improved technologies.

7.6.1 Material Heat Resistance

Resistance to thermal degradation depends almost entirely on the packaging materials, where Polyimide and silicones excel at high temperatures. A significant temperature constraint is the melting point of solder, where commonly used eutectic 63Sn37Pb alloy has a liquidus point of 183°C. The newer lead-free solders (typically 95Sn+5Ag+1Cu) melt at about 220°C, which favors use of more temperature resistant materials. In some package designs solder must reflow several times, so the materials are subjected to high temperatures repetitively.

7.6.2 Heat Conduction

Conduction of heat is typically through electrical leads to solder balls, then into copper circuits on the PCB that serve as "heat spreaders". This mode of heat transfer has a diminishing improvement after a threshold number of balls. A study was undertaken at Tessera to define the number of solder balls beyond which there was a diminishing thermal improvement. The conclusion is that ball counts beyond approximately 65 per square centimeter is unlikely to provide a significant benefit.

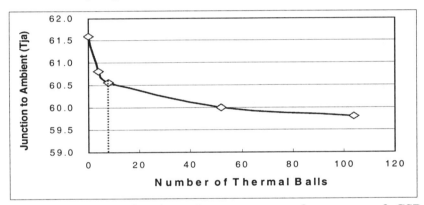

A typical application for thermal management of an array of CSP packages is the familiar metal plate across the back of several die such as used in Rambus, DDR and other memory module designs. These high-speed devices operate over 800 MHz, generating more heat than slower memory. But not all memory die are addressed at once. The objective is to provide a high thermal conductivity path from die to a shared heat sink larger than possible for individual die. This is appropriately called a "heat spreader".

Another approach involves a metal structure in the center of a stacked array to transfer heat from a small space surrounded with poor thermal conductivity material. Tessera has developed a dual purpose "Heat Spreader and Shield" which takes advantage of metal's thermal and electrical conductivity. Some typical thermal conductivity values:

Package Thermal Element	Material	Therm Conductivity Watts/(meter-degK)	Multiple of Air
PCB Conductive layer	Copper, Cu	385	>1000 X
Die Pads, Heat Spreader	Aluminum, Al	237	
Wafer Fab, most common die mat'l	Silicon, Si	124	
Shield, Heat Spreader, UBM layer	Nickel, Ni	90	
High Tin (e.g. lead free) Solder	Tin, Sn	64	
Eutectic Tin-Lead Solder Balls	63Sn-37Pb	50	
C-4 solder Balls, >90%Pb	Lead, Pb	35	
Ceramic Substrate, filler, Wafer Fab	Alum.Oxide, Al2O3	30	
Hi-Thermal Lo-Electrical Conductivity	Boron Nitride, BN	20	
Common filler for Epoxies	Calcite, CaCO3	3.3	100 X
Common filler for Silicones	Quartz, SiO2	3.0	
Thermal Interface Material	Silver filled Epoxy	2.0	
Thermal Insulation	Fiberglass	0.40	5-10 X
PCB substrate, Glass-Epoxy	FR4	0.30	
Electrical Insulator, High Temp	PTFE, Teflon	0.25	
Electrical Insulator, Rigid Forms	Nylon	0.25	
Encapsulant	Silicone Rubber	0.20	
TAB Tape substrate	Polyimide	0.12	
Die-Die attach adhesive	Epoxy	0.20	
Electrical Insulator, Flexible	PVC	0.16	
Carrier for Heat Sink Grease	Silicone Oil	0.10	
FC-72 fully fluorinated (3M Fluorinert)	C8F18, mixture	0.05	2 X
HFE-7200 fluorinated ether (3M Novec)	C4F9OC2H5	0.07	
Air	78%N2+21%O2	0.03	Air = 1X
Carbon Dioxide, 0.031% of atmosphere	CO2	0.02	
Thermal Insulation	Styrofoam	0.01	

It is clear from the chart that any kind of metal is at least two orders of magnitude better as a heat conductor than the typical materials used in packaging (e.g. silicone, FR4, polyimide). To take advantage of that 100:1 difference, a metallic heat shield or heat sink can be used against the back of the die, extending from within the stack to the perimeter where additional means of heat removal are available.

Studies were performed moving the "hot die" to the top and bottom of the stack, plus equal heat from all die. The experiment diagram and results are provided below. The conclusion is that the great majority of heat is transferred from the stack via solder balls, so the lowest die in the package runs the coolest, and becomes the location of choice for die with greatest heat dissipation.

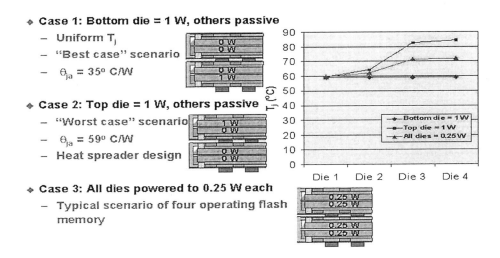

* **Case 1: Bottom die = 1 W, others passive**
  - Uniform T_j
  - "Best case" scenario
  - $\theta_{ja} = 35°$ C/W

* **Case 2: Top die = 1 W, others passive**
  - "Worst case" scenario
  - $\theta_{ja} = 59°$ C/W
  - Heat spreader design

* **Case 3: All dies powered to 0.25 W each**
  - Typical scenario of four operating flash memory

Most heat is transferred to the PCB via solder balls, so the PCB becomes a "heat sink". FR4 and similar materials are poor thermal conductors, so copper traces on the PCB become the dominant means of redistributing heat, making the amount of copper on the PCB an important design factor. Another modeling experiment was run at Tessera showing the PCB must have at least 20% copper coverage to be an effective heat sink.

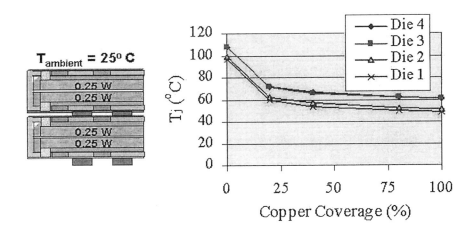

7.6.3 Convection

Convection is employed to transport heat via air or fluid flow from the device or thermal collector (e.g. heat sink or spreader) to another location for external release. Convection relies on "heat capacity" of the transfer fluid, the ability of a material to absorb and retain heat energy. A comparison of heat capacities for various materials reveals that almost anything else is better than air. Heat capacity is usually quoted related to mass, which has been converted to volume for a same-dimension comparison to air-cooling.

| Package Thermal Element | Material | <------ Heat Capacity -----> | | Multiple of Air by Volume |
		Joules /deg-gram	milliWatt-hr /deg-cm^3	
Liquid Water	H2O	4.18	1.1609	4502
Shield, Heat Spreader, UBM layer	Nickel, Ni	0.44	1.0976	4257
PCB Conductive layer	Copper, Cu	0.38	0.9528	3695
Ceramic Substrate, filler, Wafer Fab	Alum.Oxide, Al2O3	0.78	0.8537	3311
Heat Transfer Liquid, Ethylene Glycol	CH2OH-CH2OH	2.41	0.7433	2882
Die Pads, Heat Spreader	Aluminum, Al	0.90	0.6774	2627
Hi-Thermal Lo-Electrical Conductivity	Boron Nitride, BN	0.79	0.4963	1925
Wafer Fab, most common die mat'l	Silicon, Si	0.71	0.4609	1787
High Tin (e.g. lead free) Solder	Tin, Sn	0.23	0.4598	1783
C-4 solder Balls, >90%Pb	Lead, Pb	0.13	0.4022	1560
FC-72 fully fluorinated (3M Fluorinert)	C8F18, mixture	1.10	0.0045	17.3
HFE-7200 fluorinated ether (3M Novec)	C4F9OC2H5	1.22	0.0051	19.7
Freon R12,DiChloroDiFlouroMethane	CCl2F2	0.60	0.0011	4.1
Freon R22, MonoChloroDiFlouroMethane	CHClF2	0.65	0.0008	3.3
Nitrogen, 78.08% of atmosphere	Nitrogen, N2	0.74	0.0003	1.0

A previous chart shows effectiveness of metals in conducting heat, the one above shows effectiveness of metals in absorbing heat. Gases by comparison have 3 orders of magnitude less heat capacity, which gives rise to the eventual limitations of forced air cooling. Liquids, on the other hand offer a distinct advantage since they have far more mass per unit volume.

7.6.4 Liquid Cooling

Single phase liquid jacket coolers are a variation on convection coolers utilizing liquids for their higher heat capacity. Water has one of the highest heat capacities, so is commonly used in heat exchangers such as car radiators. The density of water is 1,250 times greater than air (1 gm/cc vs 1.25 g/liter for Nitrogen) which largely accounts for improved heat removal. Several water-jacket heat removal devices for electronics are on the market, including individual chip coolers and computer cases with built in coolant recirculation. [**reference 15, 16**].

A potentially useful cooling opportunity for stacked packages involves use of the solder ball separation between devices within the stack for coolant flow. Due to the high heat capacity of liquids, a relatively small flow could remove a much greater amount of heat from the stack than is possible with air. This could enable memory devices of significantly greater capacity in significantly smaller spaces. The devices and their substrates would act similarly to fins in a conventional radiator, delivering heat to the passing fluid. The following illustration shows the potentially useful "porous" nature of a CSP stacked package to fluid flow, an option not available for the folded designs.

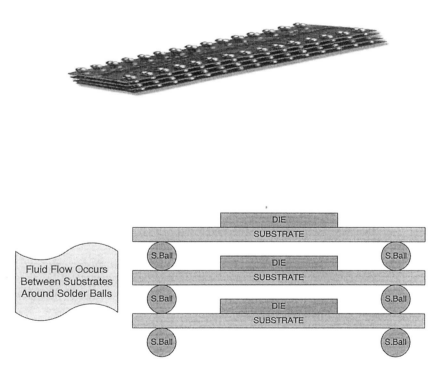

7.6.5 Phase Change Cooling

Two-phase cooling involves heat absorption via chemical phase change, where fluids cool via evaporation and liberate heat when the gas is re-liquified. The complexity of the system depends on the phase change

206

temperature of the refrigerant. Comparing heat transfer from evaporation of one cubic centimeter of 3M Flourinert FC-72 (41 milliWatt-hr) versus passage of 1 cubic centimeter of air (nitrogen at 0.26 milliWatt-hour) gaining a 10 degree temperature rise, the heat transfer ratio would be nearly 16 to 1 in favor of the refrigerant.

7.6.5.1 Spray-Cooling

Spray-Cooling is a refrigeration system developed by manufacturer ISR (Isothermal Systems Research) specifically to deal with high levels of heat dissipation from densely packaged electronics. Applications include electronics in military vehicles or high-cost floorspace where heat can be removed from a critical area and vented to a remote location without effecting the surroundings. In the case of "server farms" this method is much more efficient than using conventional air conditioning, since heat transfer via room air (with poor heat capacity) is not required, and the overall facility improvements can be less costly. In spray-cool applications coolant is pumped through an array of nozzles which direct jets of coolant directly on the sources of heat. When materials with boiling points only modestly beyond room temperature are used, there is no need for a compressor, since the gas phase will condense at room temperature given adequate heat removal at a remote radiator. The elimination of a compressor reduces size, cost, complexity, and power consumption of the installation.

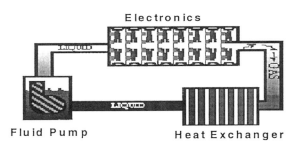

Additional benefits of closed loop cooling include a contamination free environment at the component level and reduced volume required of heat spreaders, which are currently large and growing larger for air-cooled devices. Disadvantages include a more complex sealed system with the associated plumbing and additonal long-term maintenance.

The spray cooling of stacked packages will also enable a significant size reduction for memory arrays. Conventional RDRAM, for example, requires use of heat spreaders on each side of a DIMM. A more space efficient design will include stacks of chips mounted to the PCB and the entire assembly spray-cooled. A number of sockets are eliminated and much greater memory capacity can be located in considerably smaller space. The heat generated by memory devices would be vented at a remote location, making the entire system run cooler with no load on the facility's existing air conditioning system.

7.7 Interconnect Issues

Interconnections between die can be direct (as in "Possum") where the two die have electrical and mechanical connections between them. Electrical contact methods include solder balls, wire bonds, or electrically conductive adhesives. Alternatively, two die can be interconnected via the conductors in a substrate or redistribution layer. There are benefits and limitations to each method.

7.7.1 Die-to-Die Direct Interconnect

Direct Attachment enables minimum package height, since die are separated only by an electrical and/or mechanical interconnect layer. There are three possibilities (face-to-face, back-to-back, face-to-back), all of which are in use.

7.7.1.1 Face-to-Face Die Attach

This method requires mechanically matching electrical contacts, typically solder. This construction consumes minimum Z-height, as in the case of Tessera's "Possum" or similar constructions. Each die contact must be mechanically aligned with its counterpart prior to reflow, with somewhat greater precision than surface mount PCB assembly due to the finer pad pitch.

208

Coefficient of Thermal Expansion (CTE), PPM/degC.

Material	Application	CTE
Silicon, Si	Wafer Fab, most common die mat'l	3.0
Gallium Arsenide, GaAs	Wafer Fab, RFdie	5.4
Germanium, Ge	Wafer Fab, low fwd voltage drop	6.1
Nickel, Ni	Shield, Heat Spreader, UBM layer	13.0
Gold, Au	Protective plating	14.2
Copper, Cu	PCB Conductive layer	16.6
Aluminum, Al	Die Pads, Heat Spreader	25.0
Tin, Sn	High Tin (e.g. lead free) Solder	20.0
63Sn-37Pb	Eutectic Tin-Lead Solder Balls	25.0
Lead, Pb	C-4 solder Balls, >90%Pb	29.5
Zinc, Zn	UnderBump metalization	35.0

PPM is "Parts Per Million" , e.g. mm change per mm size *10^-6

7.7.1.2 Face-to-Back

Face-to-Back attachment has both die facing the same direction, and usually involves interconnect via wire bonding. This works best when the upper die is smaller than the lower one (pyramid structure) or rotated (if rectangular) to provide access to the edge electrical pads. Center-bond die would not work in the lower die position, and may be a problem on the upper die if die size is too large, requiring wire lengths which are impractical for electrical inductance or wire sweep problems during overmolding.

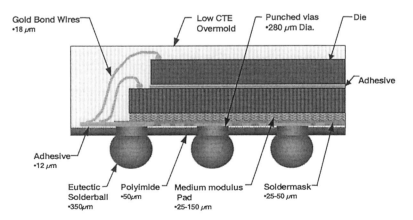

Use of same-size die requires use of a spacer to accommodate the wire bond service loop as well as multiple passes through the bonding operation and additional assembly steps to install spacer(s). Typical service loops are 3 mils high (75 micron). With thin die and resilient spacers additional care

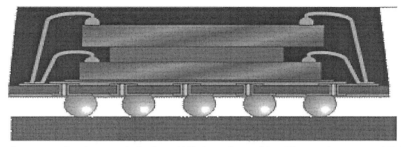

must be exercised to minimize "diving board" effects due to additional flexibility of the bonding and underlay surfaces.

7.7.1.3 Back-to-Back

Back-to-back Attachment is attractive for minimum z-height, since only a bond-line of adhesive (12-25 micron) is required between the die. This assembly provides the equivalent of a "two-faced die" available for a several bonding options. Examples include the Tessera "combo" package, or conceptually a die flip chipped to a substrate (face down) with face up die mounted on back, wire bonded and overmolded.

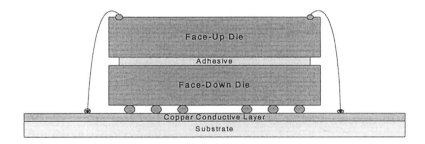

7.7.2 Redistribution Layers

Redistribution layers may be required or desirable for a number of reasons including die pad layout, pad pitch, or equivalent but mechanically different die sources. Redistribution, as the name implies, electrically connects and redistributes the die electrical pads to a desired package layout. Functionally equivalent die from different makers with unique size and layout can be made compatible by redistributing to a common format, important for second sourcing, particularly with legacy die to avoid costly wafer redesign.

7.7.2.1 Mirror Die

Mirror-image die may be required for optimum layout, since two identical die side-by-side (or on opposite sides of a substrate) may have interconnect difficulties, requiring "crossover" of wires not possible with a low cost single conductive layer substrate. An example would be memory die wired together in a mostly parallel array.

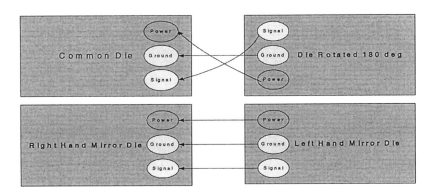

Use of a mirror-image part can simplify layout, but at the cost of dealing with two versions of the same die.

7.7.2.2 Multi-Layer Substrates

Multi-conductor substrates are commonly used in PCB designs to provide cross-over wiring, shielding, plus common layers for power and/or ground connections. For packaging of small devices, multi-layer rigid and flex substrates can be used for the same purpose. Two-metal flexible Polyimide (conductive layer on both sides of the substrate) is available from a number of suppliers. Added complexities include interconnect vias from side to side and a more rigid material (due to 2 metal layers) which provides greater resistance to folding.

A generally accepted rule of thumb is that material stiffness increases as the cube of the thickness increases (for the same material). Based only on dimensions, the 1-metal Rigid Base substrate starts out a factor of 10 stiffer than the same configuration using polyimide tape. Material property differences widen the stiffness gap.

7.8 USE OF FLEX INTERPOSER FOR COMPLEX SIP DESIGNS

The next level of electrical interconnect is to combine individual CSP packages (single or multi die) into a more complex SIP design. The previously described package types can be assembled onto a flex substrate rather than a PCB. This approach utilizes the flexibility, fine trace density and thinness of flex. The interposer provides the wirerability of a PCB. The individual packages can be tested prior to assembly. The high level of interconnect within the individual packages means most systems can be designed with 2-layer flex interposers. The structure shown below demonstrates a foldable SIP utilizing multiple package types.

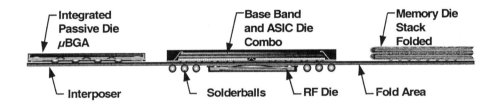

7.9 COST

Cost is usually the 1st question in consideration of a new package, followed by "what is it" and "what are the advantages". Most new packages cost more than their contemporaries, but decline over time at a rate governed by the volume. Vendors always figure out (helped along by customer pressure) how to make their materials better and less costly. The key is to have a package that has a fundamental basis for lower cost such as less material, lower material cost, more parts per batch or fewer manufacturing steps. For complex packages, one must consider the "whole picture". The extra cost in multi-die is usually offset by lower cost when used in a product. For example: fewer parts, fewer connections, smaller/cheaper PCBs, better yields, etc.

7.9.1 Redistribution Layers

Use of a substrate material adds cost but provides a number of advantages such as stress relief through compliance, a greater level of die maker independence, ability to handle a number of part at once (strip based process) and ability to accommodate die shrinks (with a tape change). The downside of a substrate is additional cost. Tape-based CSP designs have tape typically consuming 40-50% of total packaging cost. Polyimide based TAB tape is most commonly used, made by a number of vendors with a wide range of prices depending on design complexity, volume, number of metal layers, if Soldermask is required. Significant cost reduction progress has been made through vendor learning experience, additional competitors, and making designs less costly to produce. A rule of thumb has been "a penny a pin" for CSP packaging cost, but this cost has been steadily going down as CSP production volume increases. Today some CSP costs are nearing TSOP costs of under 0.5 cents per pin.

7.9.2 Folded/Stacked Package Cost Elements

The "80% rule" is usually paraphrased to the effect that "80% of the problem resides in 20% of the population". In the case of packaging, a few elements comprise the bulk of cost and are natural targets for cost reduction.

7.9.3 Tape-based SIP designs

Circuitized flex tape is the major material cost item in most designs. Folded and/or stacked designs require additional tape for the fold or the support for the peripheral solderballs. This means the cost of tape increases proportionately for these packages. A recent example of a 5-die folded and stacked SIP quoted to a major IC maker included:

Circuited flex tape	41.8% at approx. 8 cents per cm^2
Other Direct Material	6.0%
Indirect Material	6.1%
Equipment depreciation	15.6 %
Labor	11.7%
Overhead	18.9%

Single-Metal polyimide tape is the dominant configuration in use today. Certain applications require metal on both sides (2-metal tape. Two major vendors estimate volume cost for 2-metal tape at 1.5x to 2x that of 1-metal tape at high volume.

Some vendors are pursuing non-Polyimide flexible substrates as a lower cost alternative. These include LCP (Liquid Crystal Polymer) and a few other resins. There is currently not a substantial economic, availability, or technical advantage to justify a major switch from polyimide at this time.

7.9.3.1 Rigid Substrates

FR-4, BT and similar substrates have a cost advantage, with quotes about 2/3 that of polyimide. Although this is attractive, rigid substrates can't be folded, so are limited to stacked configurations. Normal FR-4 tolerances make this material less suitable for high-density SIP applications.

7.9.3.2 Solder Paste vs Solder Balls

Both methods are in use today, fully automatic solder ball application machines cost upwards of $350K, while stencil machines for paste application are considerably less. Solder ball format is also much more expensive than paste. At a typical price of 20 cents per thousand balls, the price for solder is about $300 per pound in this form factor. On a per-ball basis, 0.02 cents is cheap, but for a large part the cost becomes significant (e.g. 300 I/O would require 6 cents worth of solder balls, plus the flux). Solder paste, by comparison, is about $60 per pound, so there's a cost motivation to avoid pre-formed balls. Currently, however, the advantage of solder uniformity and reduction in inspection costs favor use of solder ball manufacturing techniques.

7.9.3.3 Assembly Cost

Although the CSP components within a SIP usually carry a cost premium, the consequent advantages of a simpler assembly offer cost savings elsewhere in the manufacturing chain.
- Fewer connections at the PCB due to interconnects within the SIP
- Smaller and simpler PCB designs are the norm often reducing the number of PCB layers.
- SIPs are "known good" via pre-testing, so PCB yields are very high
- High reliability cuts labor, rework, and warranty exposure
- Fewer parts means easier and simpler PCB rework when required

7.10   SUMMARY AND CONCLUSIONS

The folded package is an extremely versatile means to achieve high-density 3-D configurations, and can be combined with a number of divergent technologies to fit a variety of needs. Advantages include:

- Folded & Stacked provides optimum use of limited space

- More functionality in the same space, add features at no size penalty
- High system performance due to short lead lengths
- Enabling technology some medical, military, instrumentation and consumer applications
- Complete and pre-tested subsystems ready for application drop-in.
- More expensive subassembly, but cheaper/simpler final assembly
- Improved rework capability for final assembly, easier debugging
- Higher first pass yields using pre-tested sub-assemblies
- Reliability advantage when using low stress CSP technology

As with any new technology, a few risk factors exist.
- Higher initial packaging cost, offset by advantages of user simplicity and lower cost final assembly
- Heat removal from more densely packaged high wattage sources unless sophisticated cooling techniques are used
- Learning curve required for early adopters

The authors are convinced that 3-D packaging will migrate from "nice to have" to a "must have" status as more functionality is forced into ever smaller devices. There is simply no alternative to abandoning a planar world of surface mount devices in favor of utilizing the 3rd dimension. In this world folded flex will play a significant roll.

Appendix, REFERENCES and BIBLIOGRAPHY

A great deal of useful information is available on the Internet, and many references in this paper are linked to the relevant web site and area of interest within that site. Direct links will benefit those who have access to a software copy of this document. Others will have to enter site data by hand. All the links listed have public access. A few site owners may require free "registration". The hard-copy references are available from the sponsoring organizations, sometimes on-line, usually for a fee.

1. Tessera Inc. has descriptions and photographs of numerous 3-D package designs at: http://www.tessera.com/

2. Intel has been an early adopter and volume producer of advanced packages, substantial application and design information is available through their website.
   A detailed user's guide for CSP packages of many configurations at: http://www.intel.com/design/flcomp/PACKDATA/csp/29816101.pdf
   A product briefing on Intel's adoption of 3-D packages at:
   http://www.intel.com/design/flcomp/prodbref/298051.htm

3. Polyimide material descriptions available from Dupont at http://www.dupont.com/kapton/

4. Similar data from UBE covering their popular Upilex-S Polyimide films at http://www.ube.com/

5. Review of polyimide and other polymers used in packaging available from University of Southern Mississippi polymer chemistry reference encyclopedia "Macrogalleria" : http://www.psrc.usm.edu/macrog/imide.htm

6. A review of 3-D packages is provided in a paper presented by Mike Warner of Tessera to ICAPS (International Conference on Advanced Packaging) on 12 March 2002 in Reno Nevada. Sponsoring organization is IMAPS (International Microelectronics and Packaging Society) at http://www.imaps.org where proceedings can be ordered.

7. "Flex Based Multiple Die Chip-Scale Package Technology" presented to Advanced Technology Workshop on Passive Integration hosted by IMAPS in Ogonquit, Maine on 19-21 June 2002. See http://www.IMAPS.org for availability of workshop proceedings.

8. Intel paper by May and Woods, presented at IEEE 16th annual proceedings of Reliability Physics on 18 April 1978, IEEE Catalog No. 78CH1294-9PHY, Library of Congress No. 76-180194

9. J.F. Ziegler presents a historical review titled "IBM experiments in soft fails in computer electronics (1978-1994)", not currently on-line but should be available in some libraries or

from IBM. This was the feature article in IBM Journal of Research and Development, Volume 40, No.1.

10. J-STD-012, January 1996, "Implementation of Flip-Chip and Chip-Scale Technology", joint industry standard developed by EIA, IPC, JEDEC. Pages 22-23 discuss nuclear emissions levels from common materials and effects on semiconductors.., Page 75 cites additional risk for high lead solder in Flip Chip designs.

11. Pure Technologies sells lead with radioactive isotopes removed, via a Russian nuclear facility partnership. More information available at http://www.puretechnologies.com

12. Alpha particle stopping by various materials adapted from Sourcebook on Atomic Energy, by Glasstone, 2nd edition 1967 published by Van Nostrand LC#67-29947.

J.F. Ziegler also lists as a reference on IBM's website to ICRU (International Commission on Radiation Units and Measurements) publication #49 released in 1993 which has a paper "Stopping Powers and Ranges of Alpha Particles"

13. Dow Corning investigated radioactive contamination in "natural" fillers and improvements from synthetic types. For details see Ann Norris and Udo Pernisz paper at http://www.chipscalereview.com/0500/silicone24.html

A source of data on silicone shielding for radioactive contamination on semiconductors given by M.White of Bell Labs in a paper "The use of silicone RTV rubber for alpha particle protection on silicon integrated circuits" available as IEEE proceedings CH1619-6/81/0000-0043S00.75

14. ITRS (International Technology Roadmap for Semiconductors) has >200 pages of detailed information on the Semiconductor Industry available at http://public.itrs.net . This is a great resource with annually updated short term and long term forecasts for all aspects of the semiconductor industry, based on extensive surveys of the participants.

15. Examples of computer cases with built-in liquid cooling systems can be seen at http:/www.Koolance.com

16. A variety of liquid cooling solutions focused on PCs components can be seen at : http://www.coolerguys.com

17. 3M website at http://cms.3m.com/cms/US/en/2-68/iilzRFS/view.jhtml provides technical information on fluorinated dielectric and coolant materials

18. A good review article by Robert Simons "Electronics Cooling, direct liquid immersion for high power density microelectronics" is available at: http://www.electronics-cooling.com/Resources/EC_Articles/MAY96/may96_04.htm

19. ISR. (Isothermal Systems Research) offers a number of systems utilizing 2-phase cooling of electronics, see http://www.spraycool.com

Chapter 8

VALTRONIC CSP3D

Georges Rochat, Philippe Clot, Jean-François Zeberli
Valtronic

1. Introduction

For more than thirty years, Flip-Chip has brought the perfect complement needed in the Chip-on-Board (COB) technologies : an even smaller size with lower contact resistance and impedance, and a particular suitability for cost effective applications with large i/o counts. Therefore, although the subject has been widely known since the C4 process was developed, the Flip-Chip is not automatically present in today's electronic manufacture. As the world of electronics is entering the third millenium, it is not reasonable anymore that a high tech packaging firm moves forward without Flip-Chip. Some new technologies are becoming available, offering lead-free techniques, more flexible bumping, and global processes with reduced operation counts.

Since 1996, Valtronic SA has chosen to focus one of those innovative technologies using a pre-applied Non Conductive Paste (NCP) rather than a post-dispensed underfill. The present intent is to show primary aspects of such a technique, and the improvements that produced a high reliability, flexible and cost effective process. It is also to show some applications of miniaturized products.

Fig. 1 shows that current levels of miniaturization have been realized not only from component miniaturization per se, but also from functionally more complex integrated circuits, the elimination of unessential product packaging, and the reduction or elimination of traditional wire interconnects between chips and circuit board substrates. Previous Chip-On-Board and Chip-On-Chip techniques both rely on the bonding of wire leads to unpackaged IC chips that have been mounted to circuit boards with lead pads facing upward. This technique still consumes some real estate around the periphery of the chip, but less than surface mount techniques that use packaged devices. In contrast, the Flip-Chip process (introduced around 1960, but gaining popularity in the late 1980's), mounts the integrated circuit chip with pads in direct contact with the circuit board substrate, eliminating separate interconnects. The MinExt (Chip-on-Flip-Chip) and CSP3D techniques, shown in Fig. 2 are

218

further developments in miniaturization that are based on a Flip-Chip core technology. Minext mounts a smaller chip to the exposed surface of the Flip-Chip die. This smaller chip is then connected to the main circuit board through bonded wire leads. Three-dimensional Chip Scale Packaging (CSP3D) bonds Flip-Chip and surface mount components to a flexible substrate, which can be folded to achieve smaller, three-dimensional device packages. From the above informations, it appears quite likely that Flip-Chip will be a fundamental ingredient of any progress in circuit miniaturization.

Each version of Flip-Chip technology have proven suitable or preferable in certain applications. The Gold on Gold technique offers certain advantages over those that use solder or conductive adhesives, and has been shown to effectively satisfy the requirements of applications where methods using solder or conductive adhesives are technically or financially impractical.

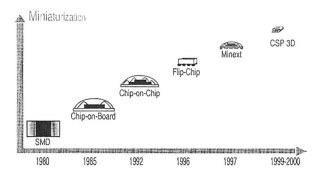

Fig. 1 : Assembly technologies, 1980-2000

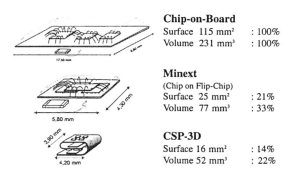

Chip-on-Board
Surface 115 mm²   : 100%
Volume 231 mm³   : 100%

Minext
(Chip on Flip-Chip)
Surface 25 mm²   : 21%
Volume 77 mm³   : 33%

CSP-3D
Surface 16 mm²   : 14%
Volume 52 mm³   : 22%

Fig. 2 : Miniaturization (%) vs Chip attachment techniques

The Flip-Chip technique reported here is used daily at Valtronic to assemble sub-miniaturized electronic devices such as hearing aids and heart defibrillators. The Fig. 3 shows a 3 dice module assembled on a multilayer FR4 substrate. The upper dice are attached "back-to-back" to the Flip-Chip die, and then wire-connected to the PWB. In this case, the excellent quality of the Flip-Chip join allows the use of wire-bonding pads that are located extremely closed to the NCP edges, typically less than 0.2 mm.

Patented technology

Fig. 3 : 3 dice module assembled with Flip-Chip and wire bonding technologies

2. Selecting a Flip-Chip technology

Three elements characterize the various Flip-Chip technologies : the bumping method, the type of substrate and the assembly process. Choosing one assembly technology rather than another is a difficult matter : as far as Flip-Chip is concerned, this choice is mainly driven by the bumping technique and the type of substrate.

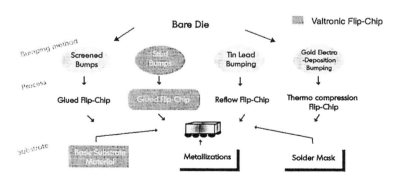

Fig. 4 : Flip-Chip technology

In our case, the gold stud-bump was elected because it offers the following advantages :
• dice can be bumped on a whole wafer as well as a single die, allowing for prototyping and production
• no additional preparation of the bonding pads is required (such as UBM) prior to bumping
• low cost bumping process that can be operated using a standard gold-bonding machine

Regarding the whole manufacturing process, it was also important that the Flip-Chip technique provides the following characteristics :
• low contact resistance gold to gold (less than 10mOhms)
• low contact impedance (compared to wire-bonding)
• compatible with the reflow process
• compatible with any organic substrates, including flex ones
• reduced assembly operations count
• compatible with production up to 1 million dice a year
• excellent reliability

For all the above reasons, the use of a non conductive adhesive appeared to be the best compromise for a flexible but high performance Flip-Chip assembly process.

3. The NCP Flip-Chip process

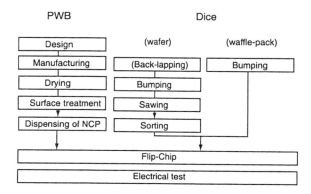

Fig. 6 : Gold Stud Bump

Fig. 5 : Flip-Chip process at a glance

Bumping

For the series production, the bumping process will preferably be completed on a whole wafer. It's also important to note that the wafer can be backlapped to less than 250 µm before bumping. The bumping consists in attaching the stud-bumps on the bonding pads of the dice using fast automatic ball-bonding machines which are now able to attach more than 15 stud-bumps per second, and making this technique a real cost-effective one. The final shape (see Fig. 6) of the stud-bump and its metallurgy are important parameters regarding the reproducibility of the process and the reliability of the final package ; the ESD protection of the wafers, at every step of the bumping process, is another essential aspect. Once the bumping is completed, the wafer must be sawed so that the dice can be sorted and stored in waffle/gel packs.The ability to bump dice one-by-one opens the door of Flip-Chip prototyping at a very low price compared to other techniques ; it's also particularly convenient for products whose package must be designed within a very short period.

Printed Wired Board (PWB)

First of all, the PWB must be designed according to specific rules to fit the NCP Flip-Chip process perfectly. Due to their manufacturing processes, and until the recent months, it has been difficult to obtain PWB for Flip-Chip assembly of dice whose bonding pads were routed at a smaller pitch than 150 microns. Hopefully, the state of the art has recently improved, and some of the new Flip-Chip products are being driven using PWB that allow the assembly of dice with a smaller pitch down to 120 microns. Starting their assembly, the PWBs are first electrically tested, cleaned and dried to eliminate any moisture within their structure. Finally, the NCP is deposited onto the Flip-Chip zone with an appropriate shape to avoid entrapped air bubbles, using either a screen-printing machine or an automatic dispenser.

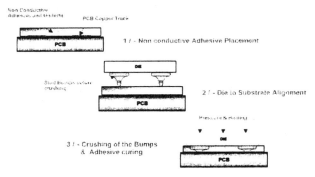

Fig. 7 : Flip-Chip process

Flip-Chip operation

Prior to joining the die and substrate, their respective connections must be aligned in order to ensure that each stud-bump of the die will make contact with its respective bonding pad on the PWB.

The die is then pressed onto the surface of the substrate, causing the crushing of the stud-bumps and their perfect adaptation to the pads of the PWB, whenever it's not absolutely planar ; the bonding pressure is mainly dependant on the i/o count of the die that are to be connected simultaneously, each bump requiring a force of 50 up to 80 grams. In the mean time, both sides of the package are heated in order to provide the thermal energy that will cure the Flip-Chip NCP.

Once the cure level is high enough to guarantee a strong bonding of the die, both the pressure and heat can be turned off. The cooling of the package to room temperature will generate an important shrinkage strength that will press the die and the PCB against each other, and apply a permanent force on each stud-bump so that the electrical contacts are guaranted.

Bonding parameters are :

Pressure/bump : 50 to 80 grams
Temperature : 150-250°C
Adhesive cure time : < 10 seconds
Alignement : ± 5 μm (± 1 μm optional)
Planarity : ± 5 μm (± 1 μm optional)

4 Industrial improvement

From the beginning of the investigations until the recent months, the industrial requirements have definitely shown that the main parameter of this assembly technique was the adhesive, and that its specification is extremely important.

Concerning the Flip-Chip on organic substrates, attention must be paid to the curing profile that must not reach temperatures that could damage the substrate. Therefore, some new chemistries have recently become available, and the «up to date» Flip-Chip adhesives are now able to cure perfectly at a fairly low temperature in less than 10 seconds. The above chart (Fig. 8) describes the minimum cure schedule for three adhesives that have been used within the main improvement steps of the Flip-Chip process.

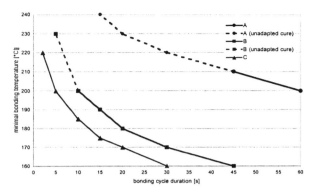

Fig. 8 : minimum cure schedule for three FC adhesives

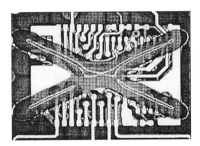

Fig. 9 : Adhesive deposition

Another important characteristic for the researched NCP is its ability to wet the surfaces of both the PWB and the die correctly. In the field of extreme miniaturization, it's also important to get a small but regular bubble-free adhesive joint around the edges of the die.

The dispensing technique (Fig. 9) is particularly efficient while optimizing the pattern of adhesive that is to be deposited on the Flip-Chip zone : both the geometry of the NCP shape and the reproducibility of the quantity are essential characteristics that produce a perfect package once the Flip-Chip operation is completed.

As far as the Flip-Chip machine is concerned, its specifications are also very strict : notably, it must be capable of applying a wide range of force and temperature while keeping accuracy on the placement of the die. Today, some equipment is becoming available that suits this kind of Flip-Chip process perfectly (Fig. 10).

Fig. 10 : Automatic Flip-Chip Machine

Fig. 11 Automatic dispensing machine

5. Performances

At the beginning of the project, it was important to know the actual potential of this technique. First of all, dice from 5x5 to 10x10mm were flipped onto a FR4 substrate : these dice provided several "4-probes" sites that offer a very accurate measure of the Flip-Chip's contact resistance ; their connections were located around the edges of the die only, at a pitch of 250 microns. These parts were then tested after 25, 50, 100, 250, 500, 750 and 1000 cycles from –40 to 125°C (air-air).

The chart (Fig. 12) shows a very stable contact resistance. No failure was observed after 1000 cycles, including the 10x10 test dice where the TCE (Thermal Coefficient of Expansion) mismatches (between the die and the FR4 PWB) were expected to create some serious problems.

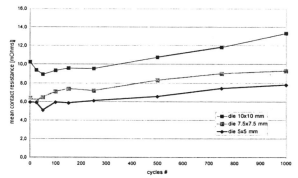

Fig. 12 : contact resistance after thermo-cycling (-40+125°C)

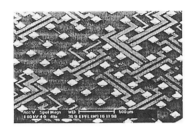

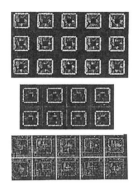

Fig. 13 : Daisy chain die with stud bumps   *Fig 14 : FR4, BT and PI test boards*

It is also important to know the behavior of the connections that are located on the center of the die ; actually, depending on the geometry of the NCP pattern that is deposited onto the substrate, some of the bumps in the center of the die will have to go through the (uncured) adhesive prior to make the contact with the pads of the PWB. A new evaluation board was then designed, using a daisy chain die that offers "bump grid array" connections and "4 probes" sites.

The NCP was dispensed so that we could observe the contact resistance of bumps that had to go through the adhesive prior making contact and bumps that could make a dry contact with the pads immediately. In both cases, these parts have shown that the contacts have nearly the same resistance, whenever the PWB's pads were covered by adhesive or not when the Flip-Chip was started. Moreover, further cross sections have confirmed that no paste was trapped between the crushed bumps and their pads, at any point of the evaluation.

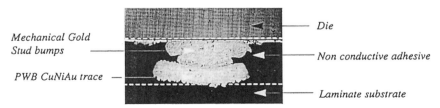

Mechanical Gold Stud bumps ——————

PWB CuNiAu trace —

———— *Die*

———— *Non conductive adhesive*

———— *Laminate substrate*

Fig. 15 : Cross section of Flip-Chip connection with possible absorption of PWB height differences

226

The following chart (Fig. 16) describes the thermal behavior of these connections located in the center and at the edge of a 4.8x4.8 test die which was assembled onto an FR4 PWB. Regarding the increasing value of the contact resistance with temperature, it was feared that each thermal cycle would damage the connection, whenever the room temperature was recovered ; on the contrary, we found there was a direct but reversible relation between the temperature and the contact resistance as is shown in the Fig 17.

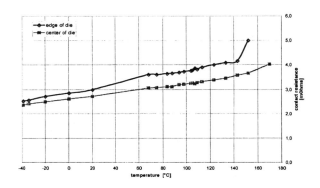

Fig. 16 : contact resistance vs temperature

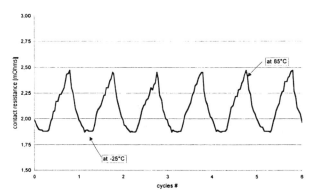

Fig. 17: contact resistance while thermocycling

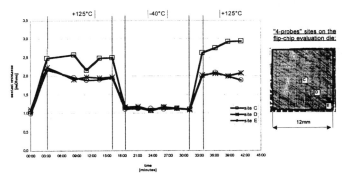

Fig. 18 : Contact resistance while thermo cycling

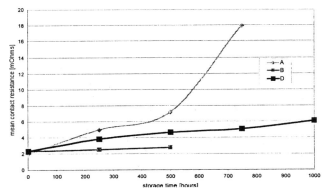

Fig. 19 : contact resistance vs humidity

Another important characteristic with this packaging process is its behavior during humidity storage. The evaluation parts were mounted using a one-layer flex polyimide PWB (thickness 50 microns) and 4.8x4.8mm daisy chain dice offering 4 sites for "4-probes" measures. One test board includes 15 dice, each of them mounted using the same NCP.

Several NCP were then tested by placing the boards at 85°C & 85% relative humidity. The contact resistances of each part have been measured after 250, 500, 750 and 1000 hours of storage.

The chart (Fig. 19) shows the behavior of the different NCP. As for cure schedule, we can see that the improvements made on the chemistries tend to produce excellent performance in humidity storage, and make this technique mature enough to be produced in large quantities.

6. Benefits of the stud-bumps on the multi layer flex substrate

As previously discussed, the gold stud-bumps are crushed onto the surface of the PWB's pads during the Flip-Chip process, creating a pure mechanical contact : its final height is automatically adapted to the planarity of the substrate.

This characteristic is particularly suitable when the Flip-Chip bonding is performed on flex substrates which are never perfectly flat :

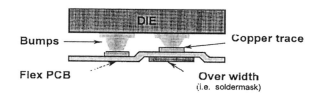

Fig. 20 : final height compensation on flex substrate

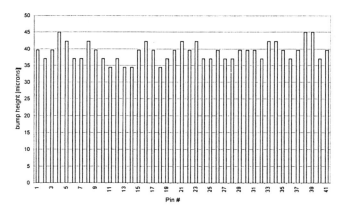

Fig. 21 : non planarity of the substrate and resulting heights of the stud bumps

7. Flip-Chip benefits/results/reliability

Benefits
• Die bumping requires no UBM (Under Bumps Metallization).
• No underfill operation needed.
• Complete electronic function integrated into the same package because SMT component implementation is possible.
• Very good resistance to thermal shocks because of independence from the TCE.
• Very low contact impedance and fast signal speed due to extremely short distances between die components and PWB.
• Die can be functionnally tested during the assembly process.

Results
An electro-mechanical contact is obtained for each bump, without soldering. The use of adhesive bonding eliminates the permanent stress which is present with soldered connections   between dice and substrate which have different TCE. Flip-Chip process performances have been verified using a Daisy Chain test chip with 256 connections. Contact resistance at To was < 5 mohm.

Thermal cycle   Humidity
-55°C +125°C 1000 cycles   85°C/85% RH   1000 hours

Reliability
To verify process reliability and product performances, many tests and cross sections have been performed.

Mechanical test :
• Vibration 10-20Hz/20-150Hz, 1,6 gr, 30 min, 3 axes, IEC 68-2-36 Test Fdb
• Drop test 1.5m on hard surface - IEC 68-2-32 Ed

Operating test :
• Temperature shock –25 + 55°C – IEC 68-2-14 Na
• Artificial sweat and salt fog 40°C 24h – IEC 68-2-52
• Damp heat test 25 and 55°C 24h 95% RH – IEC 62-2-30 Db

All samples passed, enabling the product to be qualified for several fields of application. Samples have followed the prescribed assembly process that includes reflow soldering for SMT. The reflow process has been repeated 5 to 10 times as a destructive test, followed by analysis of the variation in Flip-Chip cross section and SMT characteristics.

CSP3D Modules which withstand every Trial

Type of module & type of PCB		Die	I/O	Damp Heat without bias			Temp. Cycling	
				8.5°C/85%RH		-10°C +65°C/93%RH	-40°C +125°C	
				500h	1000h	10j	500c	1000c
CSP-3D#1	Polyimide/4 layers	3x13/15mils	50	40/40	40/40	89/90	30/31	in progress
CSP-3D#2	Polyimide/4 layers	3x11mils	50	10/10	10/10	in process	20/20	in progress
CSP-3D#3	Polyimide/4 layers	2x10mils	56	18/18	in progress	20/20	10/10	in progress
CSP-3D#4	Polyimide/2 layers	3x9/11mils	73	-	-	-	10/10	10/10

JEDEC STD 22-A101, IEC 68-2-38,   MIL STD 883 method 1010

Fig. 22

The modules covered by these tests involved four different products that were selected randomly from production batches. The table (Fig. 22) contains information essential to understanding the tests, such as the type of PCB and number of layers, the number and thickness of chips and the number of I/Os.

How to guaranty and to improve the process reliability
From the start, special attention was devoted to process testing and inspection resources. Valtronic now has environmental chambers capable of testing to the most demanding standards. Mechanical testing equipment, such as shear testers, can accurately measure the strength of stud bumps and chips. An interferometer microscope rounds out the range of apparatus by providing highly accurate 3D measurements at multiple stages of the process

The interferometer offers extreme flexibility. Its motor-driven sample holder is controlled by data acquisition software, and provides for a large range of measurement possibilities. Thus, the new machine integrated seamlessly into our Flip-Chip production line where it now provides better tracking of our assembly process at several different points:
• PCB and chip incoming check
• Monitoring bumping operations
• Measurement of dynamic response of components used for the Flip-Chips
• Surface inspection of tools produced

8. CSP 3D packaging based on Flip-Chip tehnology and products

This interesting package is the patented 3D-CSP Multi Chip Module which is based on a thin flexible PWB, 2 or 4 layers depending on the application. Flip-Chip on Flex allows for attachment of several chips on the same substrate.

Based on this principle and in order to reduce the surface required by different packages, a 3D concept has been developed. 3D packaging can accommodate multiple chips and mounting techniques including Flip-Chip and SMT. We thus are able to incorporate a complete function in a single 3D package. To develop this bending method, a full scale module was assembled with 3 Daisy Chain die. This enables reliability testing in order to test the Flip-Chip interconnections before and after bending. Finally, a cubical electronic module for hearing aid with final dimensions of 4.5 x 4 x 3mm was developed and industrialized.

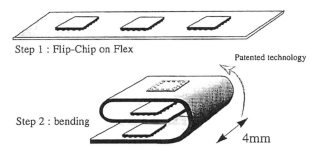

Step 1 : Flip-Chip on Flex

Patented technology

Step 2 : bending

4mm

Fig. 23 : CSP3D concept

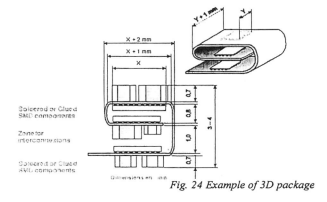

Fig. 24 Example of 3D package

Design

Mechanical design is first completed regarding radius of curves, distance between components, etc. to reach the smallest size possible.

Then, the layout is made according to electronic parameters such as power distribution, frequency and electrical test points needed for the manufacturing process. Finally, the substrate is contained in a coupon with indexing holes able to be used during the complete process. A strip includes several coupons. To accomplish the different operations of laser cutting, bending and electrical test, many dedicated tools and precise procedures are required. Some manual steps require well trained creative operators. Final test is completed using a very fine pitch bed of nails mounted in a clam shell including an X- Y table for accurate adjustment.

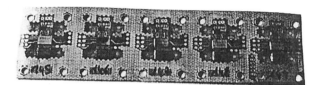

Fig. 25 : Strip for manufacturing

3D Process Flow Chart

Fig. 26 : 3D manufacturing Flow Chart

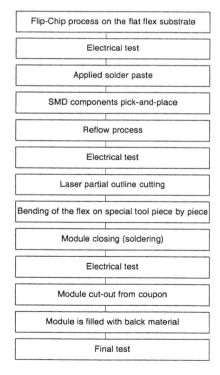

| Flip-Chip process on the flat flex substrate |
| Electrical test |
| Applied solder paste |
| SMD components pick-and-place |
| Reflow process |
| Electrical test |
| Laser partial outline cutting |
| Bending of the flex on special tool piece by piece |
| Module closing (soldering) |
| Electrical test |
| Module cut-out from coupon |
| Module is filled with balck material |
| Final test |

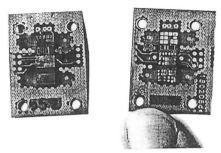

Fig. 27 : Flip-Chip side and SMD side

Fig. 28 : Steps of bending operations

Conclusion

The permanent need for more miniaturization while the number of functions is increasing is the main reason for looking to new ways to package electronic components. The use of the third dimension is a very interesting way for saving real estate and volume, but it requires a high knowledge of the flexible printed circuit boards. The combination of a flexible circuit with the FC technology is opening the door to many different applications. Due to a very low contact resistance, this technology is employed for high current applications as well as for very low power requirement. During the last 24 months, this 3-D concept was applied to more than 10 different systems and about 1 million ICs were assembled with success and high yield. In the future, more applications will be designed in order to meet the market need and requirement. As the traces and the pitch, as well as the thickness, of the flexible circuits are decreasing, the number of applications will increased as well as the level of miniaturization. This new technology needs available commercial manufacturers, and Valtronic has the technology and resources to implement thin chip and 3-D multichip packaging for European and American applications.

References

- "Gold Ball Bonding for Adhesive Flip-Chip Assembly", R. Aschenbrenner, Fraunhofer Institute Berlin
- Adhesive in Electronic 1998 - whole proceeding
- "Reliability Study of Flip-Chip on FR4 - Interconnection with ACA", R. Miessner, Fraunhofer Institute Berlin
- "Flip-Chip Joining Utilizing Gold Stub Bumps", T. Jaakola, VTT Electronics
- 0-7803-6482-1/00/$10.00 "2000 IEEE 2000 IEEE/CPMT Int'l Electronics Manufacturing Technology Symposium
- Plasma treatment for imporved bonding - A review, E.M. Liston, Branson International Corp.
- Industrial approach of a Flip-Chip method using the stud bumps with a non-conductive paste, by F. Ferrando, Adhesives 2000, Helsinki.

Chapter 9

3D PACKAGING TECHNOLOGIES: ARE FLEX BASED SOLUTIONS THE ANSWER?

Ted Tessier, VP, Advanced Applications Development
AMKOR Technology
Chandler, Arizona 85248

The 80's was the MCM era, when we all thought that multi-chip modules (MCM) were going to be the next big package innovation. MCMs were going to be the next big package innovation. MCMs were going to revolutionize packaging across the board from high end computing and military applications all the way to consumer products. Infrastructure limitations at that time, most notably IC and substrate test and yields prevented these MCMs from reaching their full potential.

During that MCM era, the Intel Corporation was an opponent of MCMs claiming that silicon integration was the way to go for the future. By the mid to late 90's, it became apparent to Intel that producing microprocessor assemblies with two or more chips were not only more cost effective than building the same function on a single chip, but provided higher yields, better performance and faster time to market. Since then, multi-die packaging has been Intel's microprocessor packaging preference.

A lot has changed in the past 20 years, Known Good Die still remains somewhat elusive for leading edge, high end integrated circuits but for FLASH, SRAM and other memory technologies major strides have been made in the availability of Known Good Die at affordable prices. Looking into the near future, this memory trend is expected to be extended towards higher end IC's with ongoing breakthroughs in wafer level test and burn-in.

Unlike in the 80's, cellular phones and other hand held appliances have largely displaced main frame computing and military electronics as the volume and technology drivers for packaging technology development in the new millennium. Major synergistic strides have been made in our generation of packaging development. High density printed wiring board technologies continue to push the limits of thin film interconnect in large panel formats. Gradual evolution in small form factor IC packaging technologies including the micro-BGA, fine pitch BGAs, wafer level CSPs and the like have led to tremendous improvements in function integration and system miniaturization.

Despite these advances, the need to integrate more and more functionality into handheld communication and computing products continues as 2.5 G and 3.0 G cellular phone requirements loom on the horizon, and in some cases have already arrived. Frankly, the industry is running out of higher density packaging options with the handheld and computing products foot print dimension restrictions.

As a result of this inevitable density bottleneck, tremendous strides are being made to develop 3D packaging solutions that can address the IC packaging requirements of portable products. A fundamental requirement of any of these emerging 3D packaging technology solutions is that they must be low cost and have a solid infrastructure to support rapid ramp-up to high volume (300 – 500 million per year) production. It is no small feat to achieve these two objectives.

Recently, INTEL made a big splash with their "Back To The Future" re-discovery of Chips-First MCM technology, originally popularized by GE in the 80's. As microprocessor design has become more and more complex, Intel is being forced to consider such "recycled" package construction technologies that might satisfy the performance requirements of their future generations of microprocessor technologies. The BBUL construction, described by Evan Davidson in chapter 2, aptly summarized Intel's vision for high performance packaging.

It remains to be seen whether Intel will be able to steer its supplier base towards providing this technology so as to create the intricate and robust infrastructure required to support it.

In the meantime, the more immediate needs of cellular phone functional integration are driving more practical, shorter-term solutions such as 3D packaging.

This chapter is intended to provide a succinct assessment of the current state of the art in IC packaging with particular attention to applications within the wireless handheld product market.

As has been discussed at length in other chapters of this book, flexible circuits and locally rigidized flex circuits have shown tremendous potential for use in a wide range of system level packaging applications. Since IC packaging, and more particularly, laminate and tape based BGAs are in widespread use in cellular phones today, BGAs remain the highest volume class of 2D applications requiring high-density substrate technologies today.

As such, these packages provide the clearest view of where packaging is likely to be headed as the migration into the 3rd dimension accelerates. A schematic cross-section comparing tape and laminate based fine pitch BGAs is shown in Figure 1.

Figure 1: Schematic cross-section of laminate and tape based fine pitch BGA packages.

Part of the motivation for my writing this chapter is to provide some balance to the optimistic accounts of 3D packaging progress being made and described in other chapters of this book. Though many of these accounts clearly highlight the promise of such packaging approaches, they often tend to downplay the challenges that remain in developing an infrastructure that is robust enough to support the mega-volumes required to support the handheld market and drive prices to where they have to where they need to be to support these applications.

9.1 2D Substrate Options

Since the starting point for the development of 3D packaging solutions inevitably lies in the infrastructure that has been established over the years to support 2D packaging applications, near term winners in the 3D packaging space will be those approaches that can most effectively leverage what is already readily available. The industry's installed capacity for tape and laminate BGA in 2002 is in excess of 1.6 billion units per year. The bulk of these volumes are largely supported by 2 layer laminate substrates and to a lesser extent single layer tape.

Table 1: High Density Flex Requirements for Emerging 2D and 3D IC Packaging Applications

Feature	HVM Ready	2003	2004	Beyond
Via Formation	Punch/Etch	Laser/Etch	Laser/Etch	Laser/Plasma
Via Fill (2 layer +)	Plated Shut/Epoxy	Plated Shut/Epoxy	Plated Shut	Plated Shut
Wirebond Pitch	100 microns	85 microns	65 microns	50 microns
Via/Pad–Min. Punch	200/325 microns	150 / 260 microns	125/225 microns	100/175 microns
Via/Pad-Min. Etch	135/260 microns	115/225 microns	100/ 200 microns	90/175 microns
Via/Pad-Min. Laser	25/125 microns	25/125 microns	20/100 microns	15/75 microns
Line / Space	25 microns	20 microns	15 microns	10 microns
Film Thickness	70 to 85 microns	70 to 85 microns	60 to 75 microns	<60 microns
# of Metal Layers	1 or 2	2	2	2
Film Material	Polyimide	Polyimide / LCP	Polyimide / LCP	Polyimide / LCP

More than 80% of the total volumes of BGA packages are based on laminate substrates because of availability and low cost. From a global perspective, there are 10 capable high density, 2 layer laminate rigid substrate suppliers for every 1 single layer tape source. In the case of comparable density multi-layer laminate and equivalent 2-layer tape substrates, the ratio is greater than 50 to 1. The net result of this discrepancy in breadth of supply is that high density 2 layer tape, the key interconnect enabler of 3D flex based packaging is currently 3 to 10X more expensive than equivalent density multi-layer laminate substrates.

Another deterrent to adoption of a flex based solution for future 3D packaging solutions are the upfront NREs required to fabricate a tape-based substrate today compared to those for a laminate substrate. This can amount to more than 10X the laminate option or in excess of 50K US$. Since this tooling is often custom, it is risky as well as costly, to tool up new designs that may never make it into high volume production. Clearly, the flex segment of the substrate supplier base needs to wake up and drive their materials and process technologies so that they can be made available at more competitive prices.

Surprisingly, the adoption of flex-friendly based process technologies have enabled laminate substrate suppliers to gradually erode the density differences between high volume laminate and tape thereby further strengthening rigid substrates position in the market at the expense of flex alternatives. A paradigm shift in the mindset of flex providers is required to allow them to compete effectively.

One of the other chapters in this book, provides a comprehensive overview of flex based technologies as well as covers the emergence of Liquid Crystal Polymers which may allow progress in tape competitiveness in future. An aggressive roadmap for multi-layer tape development is critical to keep up with the interconnect needs of emerging 2D and 3D IC packaging solutions and is shown in Table 1.

9.2 Die Stacking: Extending the BGA Juggernaut Into 3D

Additionally, the adoption of more sophisticated interconnect design methodologies like die stacking is increasingly important.

The overwhelming supply base for wirebond BGA packages has enabled the rapid deployment of stacked die packaging extensions into high volume cellular phone packaging applications. The memory requirements of 2G generation cellular phones has forced the development of this able infrastructure to the point where Known Good Die and other obstacles have been overcome over the past half decade. This has allowed the advent of 2 die stack chip scale packages (SCSPs) to be adopted widely in these handheld products [1].

Similarly, the added push of 3G cellular phones to even greater memory requirements, at a pace faster than the IC device technology could support has fuelled continued demand for more aggressive die stacking technologies. Stated more simply, the early euphoria surrounding System on a Chip solutions have largely not lived up to expectations, thereby forcing high volume applications like handheld products where survival is based on continuous innovation, to adopt more practical

System in Package (SiP) solutions. These facts have gradually driven the emergence of more complex, higher I/O SCSP applications involving integration of base band, logic and memory into a single package design.

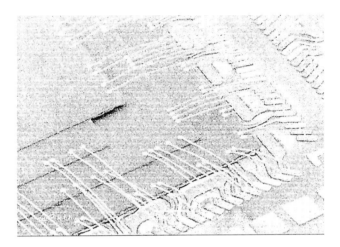

Figure 2: Scanning Electron Photomicrograph of a 4 die TSCSP Stack-up Prior to Transfer Molding

The progress made in deploying stacked die solutions to address these ever more complex 3D MCM applications is highlighted in Figure 2. This scanning electron photomicrograph shows a 1.0 mm thick, tape stacked CSP package (TSCSP) involving 4-die stacked on top of each other in a pyramidal arrangement, just prior to the over molding process. This photo shows the level of sophistication in process technology that has been enabled by the overwhelming global infrastructure in place to support high volume BGA production. This sophistication includes high volume manufacturing capable of wafer thinning and stress relief down to 50 microns, low loop height and long wire length wire bonding, die overhang in excess of 1 mm and low profile same size die stacking.

This collective process flexibility enables the mixing and matching of virtually any die combination into a very volume efficient packaging solution. The major hurdles to stacking die beyond 3 or 4 die high are related to two main issues: assembly yield impact of packaging more "Almost Known Good Die" in the same package and package substrate routing density needs associated with accommodating the interconnection of die that were never designed to play together. Based on first hand experience, it appears that the high quality of memory die can enable SCSP solutions beyond 4 die high.

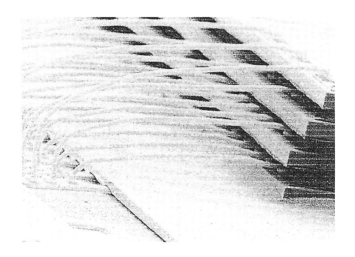

Figure 3: Low profile 4 Die Same Size Stacked CSP (SS-SCSP) Stack Up.

9.3 Fold Stack IC Packaging: The 2.5D Solution

About 3 years ago, Tessera began to offer their tape-based CSP packaging solutions to the industry. Since handheld applications are the recognized high volume driver, Tessera focused their 3D thrust in that high potential area of packaging.

Although, 2-die SCSP solutions were ramping to multi-million units per week, concerns about Known Good Die and the need to burn in the newest IC technologies forced industry leaders in 3D packaging, including INTEL and AMKOR to consider 3D packaging technologies.

Currently high volume microBGA beam lead-pitches are pretty much on indefinite hold, and have leveled out somewhere in the 100 micron pitch range. This situation is not expected to improve as most of the high volume CSP applications in the US have migrated to the substantially lower cost wire bonded fine pitch BGA packaging options.

Additionally, as described in a previous section, the supply base for high-density 2-layer tape was deemed underdeveloped by AMKOR.

We concluded that, at the present time, Tessera and its licensees have not been able to keep pace with the bond pad pitch reductions that the high volume wire bonding packaging alternative requires, therefore they are presently losing out to fine pitch BGAs that can offer higher densities for high volume, high performance applications.

Couple this with the specific beam pitch requirements of the microBGA and an unacceptable substrate supply chain supply, shortages could result.

A schematic representation of a 3 and 4 fold microBGA stack package is shown in Figure 4.

Figure 4: Schematic of a 3, 4 Die MicroBGA Fold Stack Package

From a purist standpoint, this construction is not a true 3-D construction because the circuits are routed from one chip to another through the flex circuit, which folds upon itself, resulting in the stacked array. Some call it a 2.5 D construction for that reason.

Fold stack technology, though inherently appealing at first glance, is challenged with a number of major manufacturing issues, not the least of which is the availability of difficult to source 2 layer tape substrate needed for high volume applications.

Additionally, since most foldable flex approaches are based on either Tape Automated Bonding (TAB) or the more traditional over molding and wire bonding, they are not re-workable prior to folding thereby saddling fold stack technologies with the same absolute requirements for Known Good Die as die stacking, hence no real advantage.

(Editorial Note: The European work with the flip chip and thinned dies seems to have not been available to AMKOR / INTEL in their analysis).

If thinned die are not considered, the flex thicknesses contribute additional height to the die stack as well as providing a slightly larger package footprint. If an appropriately smaller bend radius were maintained for the flex in the areas of the bends, a smaller footprint would be achieved.

When Amkor has evaluated the Tessera version of folded stack packaging solutions, it was very difficult for them to see how such an approach could cost-effectively challenge the standard extensions of stacked die packaging technology on thin rigid substrates.

As a result, we concluded that fold stack technologies as a direct competitor to die stacking will be relegated to support niche packaging applications, outside the high volume, low cost, mainstream handheld market applications.

(Editoral Note: This is in contrast to the European interest in this technology for cellular.)

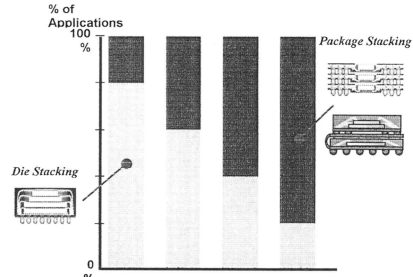

Figure 5: Die Stacking Versus Package Stacking (courtesy Amkor Technology)

It will be limited in the ability to provide z-axis connections by the high volume, fine pitch process capabilities of the day, currently a minimum pitch of 0.50 mm.

Recently, Amkor has begun high volume production of just such a rigid stackable package dubbed the extra thin CSP or etCSP^{TM}.

A schematic cross-section of this stackable package as well as a cross-section off actual parts and stacked package assemblies on a motherboard is shown in Figure 6.

Results confirming the board level reliability of such a package stack for handheld product applications were recently published elsewhere [3].

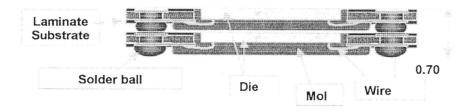

Figure 6: etCSP^{TM} Package Stacking   (courtesy Amkor Technology)

9.4 Vertical Integration: Die Stacking vs. Package Stacking

As mentioned earlier, the ability to stack die cost effectively in a non-reworkable 3D configuration is largely determined by the quality of the die that are used in it. Beyond a certain point that can vary substantially from one application and one IC provider to another, it may be necessary to consider building up the 3D stack in sub-units that can be assembled after package level burn-in and test, in a name, Stackable Packages. Figure 5 is a useful chart, which shows this imprecise trade-off graphically. For 3D applications, where the number of die in the "System in a Stack" is around 2, are truly KGD, do not require burn-in and are relatively inexpensive, SCSP wins virtually all of the time, because ultimately cost is the overriding factor in consumer product applications.

To the right hand side of the chart, as the number of die in the stack increases, as the potential risks of yield loss increases and die cost increases, the need for stackable packages emerges.

Another major driver that has been observed recently for stackable packages are situations where the end users wish to have as much flexibility as possible in the sourcing of their silicon. In this case, in order to avoid being locked into difficult single sourcing arrangements with large "total solution" IC providers, end users are entertaining sourcing chips from multiple sources in stackable packages that have been tested and if need be burned-in prior to stacking.

By adopting such a strategy, the end user is also able to mix and match devices sourced from multiple vendors depending on product type thereby overcoming one of the major limitations of stacked die packages.

Due to the significant leverage that some of the high-end users in the handheld product applications have, stackable package standardization, especially for select high volume device combinations may be achievable if enough pressure is brought to bear.

Based on the above logic, stackable package development activities are just now starting to take off, with a peak in this effort expected to occur during 2003. Widespread, high volume production and deployment of stackable packages will likely occur beginning in mid-2004.

Two primary approaches to stackable packaging are likely to emerge. One of these is based on leveraging the healthy infrastructure of rigid laminate BGAs and will involve interconnecting individual packages in the stack using solder ball connections. This will result in depopulated array packages with predominantly a fan-out design, particularly for upper packages in the stack.

 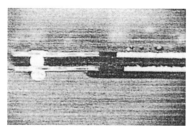

Figure 6: etCSP™ Package Stacking   (courtesy Amkor Technology)

The other approach to stackable packaging is to use a fold-over flex arm, which similarly enables z-axis connections within the stack. In the most likely scenario, an extension of tape based, wire bond, fine pitch BGAs where a fold-over flap is bonded to the top of the over molded package body, thereby providing an array configuration on which can be solder ball attached a host of other fine pitch area array packages. In the case of fold-over flex arm approach, a much higher density of z-axis interconnections corresponding to the minimum lines and spaces capabilities of the flex substrate technology and the width of the folded flap is achievable, thereby supporting higher numbers of I/Os (fan-in and fan-out) for those upper packages in the stack.

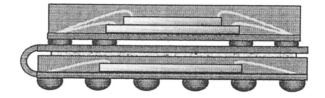

Figure 7: Fold-Over Tape BGA based System in a Stack

Figure 7 highlights a likely structure where, a high I/O, pre-tested fold stack TBGA provides an interconnection platform on which to attach a multi-die package within the System in a Stack. As became obvious while considering foldable flex die stacking options earlier in this chapter, package stacking will never be the lowest cost 3D technology, if considering only relative assembly costs.

(Ed: This is in disagreement with the expectations of Evan Davidson expressed in chapter 2.)

Total cost of ownership including additional costs associated with test and burn-in as well as other benefits that are much more difficult to quantify, benefits including sourcing flexibility must be considered. For those package-stacking applications that require significant z-axis routing density, flex based solutions may have the upper hand.

Clearly, we are just at the beginning of our evolution into the 3rd dimension in IC packaging. Several complex issues remain to be addressed in order for 3D packaging to reach its full potential and be in an adequate position to support high volume applications.

In the remaining chapters of this book, key areas of development in 3D packaging will be discussed including flex film availability electrical design; thermal enhancement and emerging substrate and other technologies that will further enable 3D packaging.

If Foldable flex and thinned silicon chips are to compete for the high volume high-end applications, the substrate costs and capabilities have to be more competitive. Perhaps LCP technology will assist that. Perhaps the flex manufacturers can lower their costs if they see volume orders, but there seems to be a need for improvement if high tech System-in-a-Package applications are to be designed for high volume.

If cellular is the driver, multiple microprocessors may not be needed for many applications. IMPS technology to provide power and ground and cross-talk control in two-sided flex may still be needed if not for cellular, then for other demanding applications.

For the most part, heat is not a major problem with present cellular applications, but may well be in the future. Some easy-to-apply heat transferring material can provide major improvement with modest construction changes.

While the following chapters address those issues, it may well be that the major application of foldable flex and thinned chips might be automotive and medical, with cellular applications and volumes a very competitive area for this technology.

9.5 Bibliography

1.   L. Smith and T. G. Tessier, "Stacked Chip Scale Packages: They're Not Just for Cell Phones Anymore!", Chip Scale Review, July 2001.

2.   Tessera, assorted publicly disclosed information.

3.   A. Yoshida, Y-H Kim, K. Ishibashi and T. Hozoji, "An Extremely Thin BGA Format Chip Scale Package and It's Board Level Reliability, 52nd ECTC, San Diego, May 2002.

Chapter 10

Availability of High Density Interconnect Flexible Circuits

E. Jan Vardaman
TechSearch International, Inc.

Dominique Numakura
DKN Research

10.1   INTRODUCTION

Flexible circuit interconnect has been an enabling technology in a wide range of applications, including automotive, industrial, military and aerospace, computer, telecommunication, consumer, and medical products. Flexible circuit or tape automated bonding (TAB) has also been a key interconnect for IC packages. Over the last few years the industry has expanded, and growth in the high density interconnect (HDI) segment of the market promises to exceed that of the traditional material in dollar value.

10.2   Definition

The industry defines HDI flex as circuits with less than 200μm pitch and/or via diameters of less than 250 μm. Applications for HDI flex include IC package substrates such as chip scale packages (CSPs) and ball grid arrays (BGAs). Computer peripherals such as disk drives and ink jet printers and liquid crystal display modules are also major applications for HDI flex. Medical products such as hearing aids, defibrillators, and ultrasound equipment, while small in terms of unit area, represent some of the highest density applications for flex circuit. *Figure 10.1.1* provides shows typical flex circuit features as a function of various applications.

248

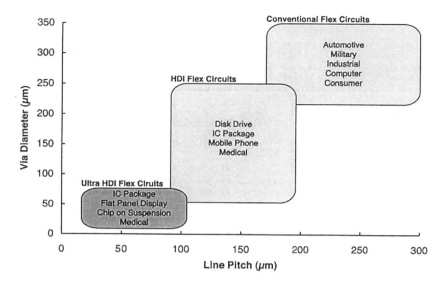

Figure 10.1.1. Flex circuit features for different applications (courtesy of TechSearch International Inc.).

10.3   High Density Flex Circuit Fabrication

Many applications, including IC package substrates, substrates for display drivers, hard disk drives, and some medical application require flex circuit with features less than 100μm pitch and microvias. As demand for these HDI applications increases, flex makers are developing new manufacturing processes with improvements in materials and equipment. While the basic process remains the same, new substrate materials such as liquid crystal polymer, adhesiveless copper clad laminates, and equipment to produce microvias have been introduced[1].

Historically, polyimide and polyester films have been the main base materials for flex circuits. Polyimide film is the primary choice because it allows high temperature processes such as soldering or wire bonding. Several new materials have been commercialized as alternatives to polyimide in the last 20 years. These materials include polysulphone, polyethylene napthalate (PEN), and liquid crystal polymer (LCP) resin. Major polyimide film suppliers include Du Pont, Ube Chemical, and Kaneka. Foster Miller has developed a proprietary processing technology for LCP that has been licensed to Kuraray of Japan and is being offered by Rogers in the United States[2]. 3M has conducted extensive evaluations of LCP films for flex circuit applications and has found that the material offers promising results[3]. Japan Gore-Tex is also providing LCP films for high density applications. Sanmina is fabricating flex circuits with LCP. In Japan, Denso has introduced a multilayer board fabricated with LCP as the dielectric and is in volume production.

More than 10 companies have commercialized adhesiveless clad laminates for HDI flex circuit in the last decade. The use of flex circuit without an adhesive layer is expected to grow as many companies seek finer features sizes and want to eliminate the low heat resistance of the adhesive layer. There are three types of adhesiveless copper clad laminates—cast, sputtering/plating, and lamination. Nippon Steel Chemical Corporation has a major share of the market as a result of its large capacity—the company uses a cast process. Additional companies with large capacity include Gould Electronics, Mitsui Chemicals, Sheldahl, and 3M.

Punching or mechanically drilling is the most common form of via fabrication in flex circuit production. Microvia formation includes photovia, laser, plasma, or other means to fabricate small diameter holes and to produce small capture pads. Many manufacturers are increasingly turning to lasers for microvia generation. Choices include carbon dioxide, excimer, or UV-YAG. Laser drilling is reported to be able to achieve minimum via diameters as small as 10 µm. Plasma processing is also used in production for medical applications.

Fine lines and spaces are common for TAB tape makers, but traditional flex circuit makers must catch up. Most TAB tape manufacturers are capable of producing circuits with 25µm lines and spaces, but only a few traditional flex circuit manufacturers can fabricate 40µm lines and spaces on 18µm copper with reel-to-reel processing lines. Material requirements play an important role in the ability to produce fine patterned circuits. Many companies are turning to adhesiveless copper clad laminates to achieve their goals. Major merchant suppliers include Nippon Steel Chemical Corporation, Mitsui Chemicals, and Gould Electronics among others. Companies such as 3M have large internally production capacity for adhesiveless flex.

Leading edge flex circuit manufactures have been developing new manufacturing processes to fabricate 15µm line and spaces on 5µm copper conductors. For high density traces smaller than 40µm pitch, dry film type resists do not work well and liquid resist is the major alternative. Additive processes are an alternative the etching process and can be used to fabricate fine lines smaller than 20µm pitch. Several manufacturers are capable of producing 5µm line/space with the additive process. These applications are for medical or industrial products.

Laser direct imaging (LDI) is a new concept in equipment that generates high density fine lines. More than three system manufacturers have commercialized new equipment for this process. Today, their fine line capabilities allow 25 to 40µm spaces. It is expected to take several years before these systems have the same capabilities as collimated light source exposure systems.

Many HDI applications require flying leads as standard construction for high density terminations. Fine flying leads with widths smaller than 75 micron and Ni/Au plating function as bonding wire. Wireless suspension or flex on suspension for disk drives require fine pitch flying leads. Traditional flex circuit manufacturers supply 100µm pitch flying leads in volume. TAB tape manufacturers supply even finer pitch (50 microns), but refer to flying leads as inner leads.

The coverlay functions as the mechanical protector of fragile conductors on the flex circuit and also serves as a solder mask during assembly. Typically, the same film used as the substrate is selected as the base material for the coverlay. Increasingly, companies are using photoimageable materials—liquid and dry film. There are advantages and disadvantages to both liquid and dry film including cost, application techniques, and chemical resistance. While dry film is easier to process with vacuum laminators that are readily available, liquid materials can provide a low-cost coverlay for large volume production with improved performance. More than 10 manufacturers have developed different types of photoimageable coverlay for HDI flex applications. Additional players are expected to emerge.

10.4   High Density Flex Circuit Applications

HDI flex circuit producers are located in many geographic regions of the world—North America, Japan, Taiwan, Hong Kong, and Europe. These companies supply an increasing number of applications for high density flex—including IC packages, liquid crystal display (LCD) modules, computers and peripherals, and medical applications. Many of these applications required that the flex circuit be folded.

10.4.1 IC Packages and LCD Modules

IC packages are one of the fastest growing applications for HDI flex. Historically, TAB tape has been used for LCD drivers and processors used in laptop computers. Packages using TAB tape are often referred to as tape carrier packages (TCPs). TAB tape is also increasingly used as the substrate for IC packages such as ball grid array (BGA) and chip scale packages (CSPs)[4]. Despite their common material properties, construction, and fabrication methods, TAB tape and flex manufacturers typically have been considered separate industry segments. With many suppliers targeting the same applications, boundary distinctions are blurring and competition is expected to increase.

Shrinking feature sizes are the trend for flex circuit (also known as TAB tape) used in applications such as LCD drivers, CSPs, and BGAs. *Figure 10.1.2* shows an example of TAB tape for a Tessear μBGA®. Lines and spaces for some of these products are 25 μm or less. Circuits with two metal layers used in IC packages now require 25μm vias. Feature sizes for HDI flex applications range from medical products with 50μm lines and spaces and 50μm vias with 100μm capture pads, to ink jet printer cartridges with 75μm lines and spaces. With increased product miniaturization and the growing use of implantable devices, feature sizes will continue to shrink for medical applications.

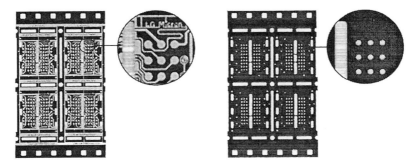

Figure 10.1.2. TAB tape for a Tessera μBGA® (courtesy of LG Micron).

Unlike many flex circuit publications, this chapter discusses both TAB tape and traditional flex circuit manufacturers. Despite their common material properties, construction, and fabrication methods, tape and flex manufacturers typically have been considered separate industry segments (see *Table 10.1.1*). With many suppliers targeting the same applications, boundary distinctions are blurring and competition is expected to increase.

Table 10.1.1. Comparison between Flex Circuits and TAB Tape

	HDI Flex Circuits	TAB Tape
Standard widths (mm)	250, 300, 500, 610	35, 48, 70, 96,105, 158, 165, 150, 305
Typical application size (mm)	300 x 300	24 x 24
	250 x 250	35 x 35,
	150 x 150	35 x 70
	100 x 100	48 x 48
		70 x 70
		105 x 105
Largest circuit/panel size (mm)	610 x 915	105 x 105
Minimum pitch for volume production (μm)	50	40
Minimum pitch for limited production (μm)	10	30
Flying leads construction	Possible	Standard
	100μm pitch	50μm pitch
Double sided with through-hole	Standard	Possible
Coverlay/solder mask	Standard	Standard
Microbump construction	Possible	Possible
Multilayer construction	Possible	Difficult
Standard shipping format	Single Units	Reel
	Sheets	Strip

10.4.2 Computers and Computer Peripherals

Computer peripherals include hard disk drives (HDDs), printers, and scanners. These products account for a large percentage of the high density flex market and represent technology drivers.

Wireless suspension or flex on suspension (FOS) in hard disk drives (HDDs) is one of the major application areas. The typical head suspension is single sided with flying leads and 70 to 100μm pitch. Wireless suspension systems for hard disk drives have been the major driver for the high density flex circuit. Traditional flex circuit manufacturers could not meet the requirements of disk drive manufacturers in the early stage, several new manufacturers developed new concepts. These companies were able to fabricate flex circuits with 40 to 50μm line/space on stainless steel suspensions. Seagate developed another technology in which flex circuits with 50μm line/space were bonded on stainless steel suspension. The wireless suspensions generated an additional demand for high density flex circuits for head wiring repair—interposer flex circuits (single sided) and control flex circuits (double sided).

The transition from traditional wiring to wireless suspension was completed in 1999. The new flex circuit technology for this application is chip on suspension. This technology will feature double-sided flex circuits and finer lines and spaces, sometimes smaller than 25 μm. Flip chip devices are used in this application.

High density flex circuit used in ink jet printer cartridges remains one of the largest segments in the computer peripheral and office equipment sectors. The typical flex circuit for an ink jet cartridge has 75μm lines and spaces and 150μm inner lead pitch.

Some connectors in high-end computers such as servers use HDI flex. The dendrite connector is one example (see *Figure 10.1.3*). Two-metal layer TAB tape used in this application has 50μm lines and 75μm spaces with 100μm vias.

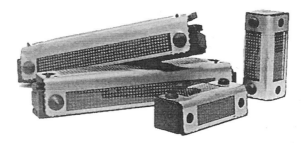

Figure 10.1.3. CPOP connector (courtesy of IFT).

10.4.3 Medical Applications

Medical application use flexible circuits to satisfy a diversity of requirements for a wide range of products. Applications include defibrillators, hearing aids, pacemakers, and sensors, as well as medical imaging and monitoring equipment. Medical products typically use flex circuit with 50μm lines and spaces and 50μm vias with 100μm capture pads. With increased product miniaturization and the growing use of implantable devices, feature sizes will continue to shrink for medical applications[5].

10.5   Flexible Circuit Suppliers

Flex circuit production takes place all around the world. Japan continues to be an important player with as much as 50 percent of the world's production, but recent expansion in Taiwan, Hong Kong, and Korea has created new sources of supply. North American and European suppliers still provide flex circuit for a variety of applications.

Major Japanese suppliers include Mektron, Sumitomo Electric, Fujikura, Nitto Denko, Sumitomo Bakelite, Sony Chemicals, Sakai Electronics, Sharp, Maruwa Seisakuksho, Okuno Seisakusho, Cosmo Electronics, Fuji Kiko Denshi, Santa Light Metal, Hitachi Chemical, and Taiyo Kogyo. TAB tape manufacturers such as Mitsui Mining and Smelting, Shindo Denshi, Hitachi Cable, Casio, CMK, Sumitomo Metal Mining, and Toppan should also be counted as flex circuit suppliers.

There are more than 20 flex circuit makers in Taiwan. These companies include Career Tech, Complex Micro Interconnection, Dowlentec, Flexium Interconnect, Hajime, Hsin-Sun Technology International, Lei An Electronics, Mutual-TEK Industries, Neoflex Technology, OFlex Technology, Picotec International, Pucka Industrial, Sea Star Electronic, Sunflex Tech, Tai Flex Scientific, Uniflex Technology, Vertex Precision Electronics, Ya Hsin Industrial, and WUS.

Flex circuit makers in Hong Kong include Asia Membrane Keyboard Switch, Advanced Tech. Holding, Compass Technology, Hong Yuen Electronics, Pantage, and Sam Sun Development. Companies in China include China National Aero-Tech, Suzhou Flexible PCB Electric, Suzhou Instrument Elements, and U.S.-owned Parlex. Korean flex circuit makers include Acqutek (associated with Amkor/Anam), Korea Circuits, LG Micron, New Flex Technology, Samsung Electro-Mechanics, San Yang, and SI Flex. Flex circuit manufacturers in Singapore include MFS Technology (formerly called M-Flex) and U.S.-owned 3M.

Major flex circuit producers in the United States include Innovex, M-Flex, Parlex, and 3M. Sheldahl has historically been a major supplier, but has recently filed Chapter 11. There are dozens of small flex circuit makers in the United States. Several have excellent fine line manufacturing capabilities, including Dynamic Research Corporation. New players in the United States include the Tessera spin-off Flex21chip and start-up Interconnect Systems, Inc. (ISI). European flex circuit makers include several small companies such as Cicorel, Dyconex, and FCI Microelectronics. Larger companies include Freudenberg Mektec and Flextronic.

References

1. Dominique Numakura, "Introduction of High Density Flexible Circuits," Nikkan Kogyo Shinbun, Tokyo, Japan, December 1998.

2. Brian Farrell and Michael St. Lawrence, "The Processing of Liquid Crystalline Polymer Printed Circuits," 2002 Electronic Components and Technology Conference, May 2002, pp. 667-671.

3. Terry Hayden, New Liquid Crystal Polymer (LCP) Flex Circuits to Meet Demanding Reliability and End-Use Applications Requirements," 2002 International Conference on Advanced Packaging and Systems, pp. 116-122.

4. E. Jan Vardaman, "High Density Interconnect: Flex Circuits for IC Packages," Proceedings: 6th Annual National Conference on Flexible Circuits: A Technology Comes of Age, June 8-9, 2000, p. 144.

5. E. Jan Vardaman, Flexible Circuits for High Density Applications, TechSearch International, Inc. August 2000.

Chapter 11

RECENT ADVANCEMENTS IN FLEX CIRCUIT TECHNOLOGY USING LIQUID CRYSTAL POLYMER SUBSTRATES

Dr. Rui Yang and Dr. Terry F. Hayden
3M Microinterconnect Systems Division
Austin, Texas U.S.A.

11.1   INTRODUCTION

The multichip, foldable package technology requires high-density and compact metallized substrates, thus offering an ideal new set of applications for flex circuits. Fine-featured flex circuits have traditionally been made with polyimide films (25-76 micron typical thickness), but a new process has recently been developed with an alternative substrate -- liquid crystal polymer (LCP). Fine lines and spaces and dielectric features for advanced circuit designs equivalent to cutting-edge polyimide circuit features and not previously attainable with LCP substrates can now be produced on LCP. Flex circuits made from LCP use an adhesiveless, roll-to-roll process flow similar to that used for polyimide substrates, including vacuum metallization, additive plating and chemical etching of the dielectric.

LCP is a thermoplastic, aromatic polyester with low moisture absorption and better electrical properties than polyimide with negligible moisture effects on these properties [1]. While polyimide can be used for foldable flex multichip packaging, LCP is expected to extend the use of flex circuits to more extreme end-use environments and higher frequencies [2].

This chapter reviews the different LCP molecular structures, bare film properties and metallization methods to make flex circuits, including recent process technology developments that will enable higher-density, finer-featured designs. This is followed by LCP circuit test data on dimensional stability, reliability and compatibility with other materials and assembly processes that is making LCP flex circuits attractive for use in a variety of applications. Indeed, the thermoplastic nature of LCP makes possible the option of cohesively bonding an LCP film "coverlayer" to an LCP circuit. This bond achieves better moisture protection than is possible with polyimide in foldable flex multichip packaging, including the extension from one to two-metal-layer (1ML to 2ML) flex circuits, is summarized.

2 LCP SUBSTRATE FILM

LCP films developed recently with advanced processing techniques have resulted in an alternative substrate film to traditional polyimide for flexible circuits. This section provides a brief overview of LCP film and property improvements, outlining some unique characteristics of suitable "circuit-grade" LCP films.

2.1 LCP Film Development

Low molecular weight liquid crystalline compounds were discovered almost a century ago [3]. Liquid crystalline polymers have attained their prominence during the last three decades [4]. Specific structural features of liquid crystalline polymers usually involve a succession of para-orientated ring structures to give a stiff molecular chain with a high axial ratio (i.e., ratio of length of molecule to its width).

The rigid, long, molecular chains tend to line up in a preferred direction (machine direction or MD) during the melt stage or extrusion process [5,6], resulting in material properties with a strong orientation dependence. A uniaxial orientation is ideal to fabricate strong fibers [7,8], but is undesirable for many applications.

Foster-Miller developed a counter-rotating die technology, which moves one or both surfaces of the die transversely with respect to one another, producing a biaxial film with balanced material properties in two dimensions. Unfortunately, this processing results in curl of the web due to residual stresses in the film. To balance the residual stresses, a tri-rotating film extrusion die was developed to produce film orientations symmetrical about the midplane [9,10]. Later, Japan Gore-Tex Inc. and Kuraray Co., Ltd. also developed processes for manufacturing LCP film. Their processes differ from Foster-Miller's, but the films produced by both suppliers have very balanced properties [11].

Aside from anisotropic properties, early LCP films also exhibited many quality problems including curling, brittleness, surface roughness, embedded particulates, and discoloration, which resulted in unacceptable roll-to-roll circuit processing along with poor circuit performance and reliability.

Another important improvement for LCP film properties and quality has been the application of serial high temperature processes after film extrusion [12]. For example, the coefficient of thermal expansion can be adjusted; film flatness can be improved; and heat resistance and mechanical strength can be increased. The combination of new film process technologies and high temperature techniques has resulted in tremendous progress in improving LCP film properties and eliminated almost all early quality problems. As a result, LCP film has become an advanced alternative substrate to traditional polyimide film in flexible circuit construction.

2.2 Molecular Structure

LCP films are made by both existing as well as new proprietary thermoplastic processing methods, beginning with pelletized LCP resin. LCP resin is broadly manufactured around the world. The thermoplastic resins are typically classified into three types according to their heat resistance. The molecular structure of each type is shown in Figure 1.

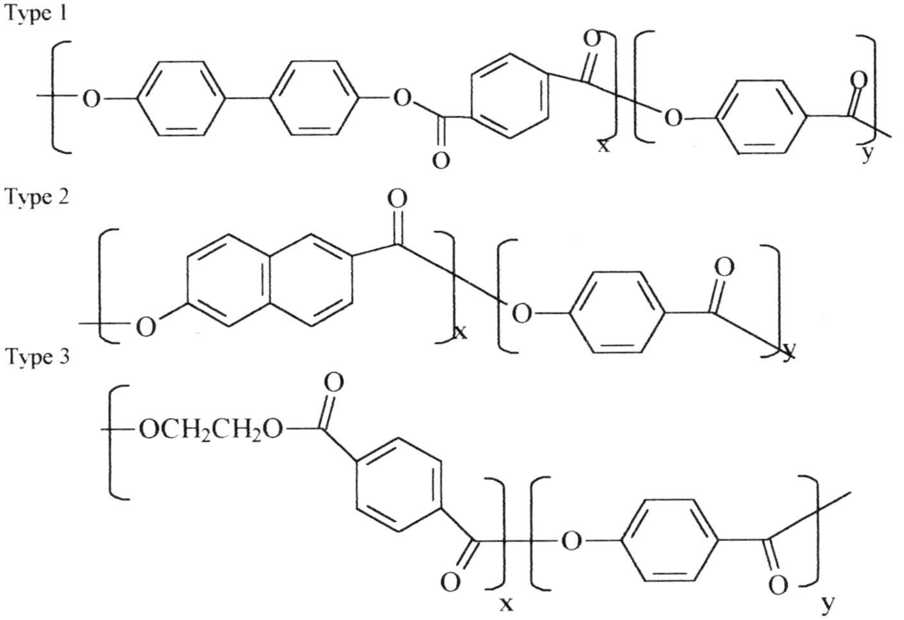

Type 1

Type 2

Type 3

Figure 1. Three types of LCP molecular structures

Type 3 LCP can be hydrolyzed because of the presence of aliphatic groups in its backbone when it exposed to water at elevated temperature. Many circuit processes and packaging assemblies undergo wet processes at high temperature. Therefore, this type of LCP film is not suitable for use as a flexible circuit substrate.

On the other hand, Type 1 and 2 LCP resins are aromatic polyesters with rigid-rod molecular structures having a much higher heat resistance than the Type 3 resin. Type 1 has a higher heat resistance than Type 2 LCP with melt temperature ranges of 300-350°C and 200-250°C, respectively. Both

LCP films from Type 1 and 2 resins have been used to make flexible circuits and are suitable for electronic packaging applications.

LCP film with high temperature resistance is ideal for final film properties, but as mentioned in the previous section, it requires some difficult high-temperature film processing steps to enhance the melt temperature by modifying the crystallinity and imparting a high degree of molecular orientation to the LCP film.

Another approach is to create co-extruded LCP polymer blends with materials such as polyethylene and polypropylene [13]. The heat resistance of these so-called "Type 4" LCP films is normally comparable to Type 2 resins, but they have advantageous processing temperatures compared to Type 3 resins. The Type 4 film has not been used in flexible circuit manufacturing to date.

2.3 LCP Film Properties

2.3.1 General Properties

Important film property similarities and differences are shown in Table 1 for LCP and two typical polyimides. As discussed in the section 2.2, the main difference between Type 1 and 2 is the heat resistance. Figure 2 shows dynamic mechnical analysis data measured from both Type 1 and 2 LCP films along with a typical polyimide film. The Type 1 LCP and polyimide film are stable and maintain most of their mechanical strength without softening at the temperature up to 300°C, but Type 2 LCP film has

Table 1. Typical property values of 50 micron thick films in the machine direction. Listed data are for Type 1 LCP films although Type 2 LCP has similar values to Type 1.

Property	PI 1	PI 2	LCP	Test method
Tensile strength (kpsi)	50	42	15	ASTM D882, 64T
Elongation (%)	60	40	16	ASTM D882, 64T
Young's modulus (kpsi)	800	825	700	ASTM D882, 64T
Prop. Tear strength (gm)	26.2	17.5	15.4	ASTM D1922-00A
Heat shrinkage (200°C, %)	0.08	0.04	0.04	ASTM D2732
CTE (ppm/deg.C)	13	14	18	ASTM D696, 44
Moisture absorption (%)	2.4	2.0	0.1	ASTM D570, 63
CHE (ppm/%RH)	9	8	2	ASTM D570
Moisture transmission (gm/sq cm/day)	4.2	3.8	0.4	ASTM F1249
Dielectric constant	3.3	3.1	3.0	ASTM D150
Dissipation factor	0.005	0.005	0.003	ASTM D150
Dielectric strength (KV/mil)	6.0	5.7	5.9	ASTM D149

Note: The actual values of these film properties may be slightly different due to the variation of film manufacturing and test methods.

two transition temperatures and begins to lose its mechanical strength at a relatively low temperature. However, post-extrusion high temperature treatments of the Type 2 film can improve the modulus and heat resistance over that of the untreated films.

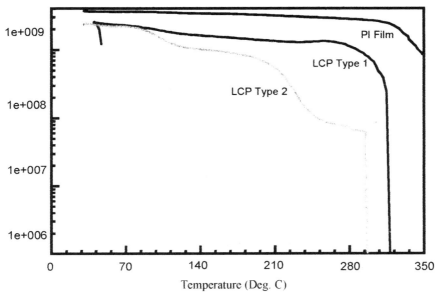

Figure 2. Comparison of Dynamic Mechanical Analysis (DMA) Modulus data of one polyimide film and two types of LCP films. DMA provides a measure of the glass transition temperature and mechanical modulus behavior with temperature.

2.3.2 Moisture Properties

A high-moisture content in a substrate film, as exemplified by the moisture absorption and coefficient of hydroscopic expansion (CHE) properties, can result in significant processing difficulties, that change circuit performance and reduce circuit reliability. For example, factory processes and environment can cause unwanted flatness (or curl) and dimensional (or registration) changes in flex circuits from built up internal stresses at the polymer-copper or polymer-secondary material (e.g., soldermask) interfaces. These can halt operations and lower yields and quality in both manufacturing and assembly of flex circuit products.

Besides affecting curl and dimensional stability, moisture absorbed during manufacturing, testing, or field operations can also cause an increase in both the dielectric constant and dielectric loss, which can change the characteristic impedance and propagation delay of the flex circuits. In addition, evaporation of the water in the circuit film at high temperature can cause blistering, delamination and metal corrosion.

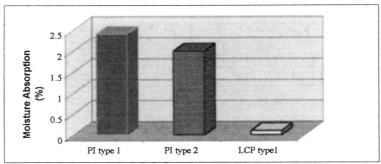

Figure 3a. Moisture absorption comparison between LCP and polyimide films of 50 micron thickness

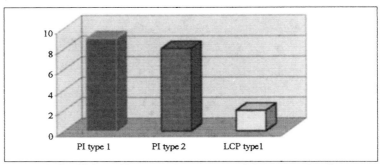

Figure 3b. Coefficient of hydroscopic expansion (CHE) comparison between LCP and polyimide films of 50 micron thickness

Compared to typical polyimide, which is prone to curling, LCP film has much less moisture absorption and CHE, which are an order and a half-order of magnitude less than polyimide respectively (see Figure 3). Therefore, LCP circuits are much less sensitive to environmental changes.

2.4 Comparison of LCP and Polyimide Films

Currently, polyimide film is the primary substrate choice for use in flexible circuit constructions. Polyimide film has a number of excellent properties that has led to its current market dominance. However, polyimide film also has some inherent weaknesses that have limited its capabilities in meeting requirements of certain applications. Key among these are the properties relate to moisture uptake. In light of this, the advantages of the LCP film are summarized in the following.

1) Electrical Properties - LCP films have a lower dielectric constant and dissipation factor than polyimide films, and these properties remain relatively constant as surrounding relative humidity increases. The effective functional frequency range of circuits constructed from LCP film ranges from 1KHz up to 45 GHz [14]. LCP films potentially can have metal trace signal

lines placed closer together with minimal crosstalk, due to these better electrical properties when compared to polyimide.

2) Reliability - LCP films experience much less of the water uptake issues associated with high temperature, including evaporation, blistering, delamination, metal corrosion and via poisoning (oxidation of the metal forming vertical connections).

3) Dimensional Stability and Curl - LCP films maintain their dimensions in high relative-humidity environments, resulting in a dimensionally stable and a flat circuit construction with little internal stress.

4) Chemical Resistance - LCP films exhibit significantly higher levels of solvent and chemical resistance than polyimide films, rendering them a good choice for applications with challenging aggressive environments.

In addition to the moisture and chemical resistance advantages, LCP film also offers additional unique properties, which include:

1) Thermoplasticity - Unlike polyimide, LCP film is a thermoplastic material: does not set or cure when heated. LCP can be softened to a flowable state when heated and hardens when cooled. Taking advantage of the flow of the thermoplastic materials when melted, LCP can be thermally laminated directly to copper foil, stiffeners, and even itself, opening up whole new avenues for circuit design and construction.

2) Adjustable coefficient of thermal expansion or CTE - Unlike polyimide, the magnitude of CTE of LCP film is adjustable through the use of thermal treatment processes. The in-plane CTEs (machine and transverse directions, MD and TD) can be balanced and matched to copper by applying the appropriate thermal treatment. Furthermore, LCP circuits show much reduced thermal expansion effects, which results in flatter circuits.

3) Unique Internal Stress Relief Mechanism - Unlike polyimide, studies have shown that LCP film undergoes a rapid ester interchange between chains at the interface at high temperature. The interchain trans-esterification relieves the interfacial stress and limits the interfacial stress build-up [15]. Thus, the flatness of LCP circuits is excellent and showing much less thermal effect throughout the service life of the circuit.

Clearly, LCP films exhibit some unique characteristics and advantages that allow them to excel in certain applications. However, LCP films do have some disadvantages that must be carefully taken into account when evaluating their use for flexible circuit applications. Key among these are the following:

1) Mechanical Strength - LCP films have lower mechanical strength and toughness than polyimide film, which can be a disadvantage in both flexible circuit manufacturing and in end-use requirements.

2) Fracture Toughness - LCP films may have a relatively poor notch sensitivity.

3) Temperature Resistance above 280°C - LCP films may not perform well in continuous operation at elevated temperatures above 280°C. Fortunately, however, this is not expected to limit many electronic assembly

applications, because most involve eutectic tin-lead or lead-free solders at which require lower processing temperatures.

In summary, LCP films can be exploited for their outstanding moisture uptake, thermoplasticity and other properties. The result is an exciting new alternative electronic substrate that can surpass the world-class performance of polyimide in certain applications.

2.5 LCP Film Availability

The beneficial properties of LCP films have been known for many years, but the poor film quality caused by processing difficulties has prevented the use of LCP in flexible circuit applications in the past.

Now LCP films are available commercially from a few suppliers, who make exceptional quality electronic-substrate-grade multiaxial LCP films with controlled molecular orientation and balanced film properties. Depending upon a number of factors, including the circuit-making process to be employed and the end-user requirements, both Type 1 and Type 2 LCP films are available.

Certainly, LCP film cost is an important factor in the decision making for selecting a suitable substrate. LCP film price has been steadily dropping because of the new process technologies and production volume increases. The price-performance ratio for end-product is getting very attractive, and the price is becoming very competitive with the current polyimide price.

3 LCP FLEX SUBSTRATES – HOW BEST TO INCORPORATE METAL WITH LCP FILMS

Although LCP flexible circuits have been available for some time, they have not found widespread use in electronic assemblies. This is partially due to the lack of circuit process capability, low performance and low reliability as the result of conventional processing techniques using LCP-metal laminate substrate films. In this chapter, substrate or substrate film is defined as the polymer-metal building block for flex circuits. Substrate film provides the essential foundation in the process to build flexible circuits. It is very important to choose the right type of substrate and to understand its performance. One of the key technical challenges for preparing LCP substrate has been vacuum-metallization of LCP film.

3.1 Adhesive Laminates

LCP flexible circuits have been made using three types of substrates: adhesive laminates, adhesiveless laminates and vacuum-metallized substrates. Adhesive laminates are constructed by laminating LCP film to a copper foil with an adhesive layer in between. In the early days of the electronics industry, the adhesive laminates were developed to provide designers and fabricators with interconnections

The use of an adhesive layer in the substrate may create some process and performance issues. For example, the adhesive layer materials may not have same patterning capability as the base film. Flexibility, chemical resistance, thermal resistance, moisture absorption, and electrical performance may also be different from the base film. As electronic packaging have become more sophisticated (higher densities, smaller sizes, etc.), the performance requirements have begun forcing the elimination of the adhesive layer.

3.2 Adhesiveless Laminates

Adhesiveless substrates offer a variety of advantages over adhesive laminates. Among these advantages are thinness, light weight, flexibility, thermal stability, laser and plasma patternability, and possibility for incorporation of very thin copper (a few micron). Unlike polyimide, LCP can be thermally laminated directly to copper foil in a continuous process without using an adhesive layer in between [16,17].

Good metal adhesion is however a challenge for the adhesive-free LCP laminates. In order to improve adhesion, a rough-surface copper foil having a certain tooth profile is needed for LCP lamination. Peel strength from the rough copper surface laminates can be significantly increased, which is the result of the increased surface areas and deep penetration of the copper teeth.

However, if the surface is too rough, circuit processing and product reliability can be affected. For example, it is difficult to remove all copper teeth penetrated into LCP film during the copper etching process. In order to ensure removal of the copper between traces, circuits are often overetched, resulting in undesirable trace sidewall profiles and small dimensions at the top and bottom (undercutting) of the traces, all of which can affect yield and reliability, especially for finer circuit pitches.

Figure 4 provides an example of two copper nodules remaining at the LCP surface after the copper etching process is completed. The remaining metal is a result of copper nodules being broken from the foil surface and are buried in the LCP due to the depth of its penetration. The selection of proper copper foil is critical to meet the requirements for good adhesion and process capability.

Another disadvantage associated with the LCP laminates is that copper thickness is limited to the availability of the copper foil used to make the laminates. Thinner copper foils are expensive and are difficult to handle without a supporting film.

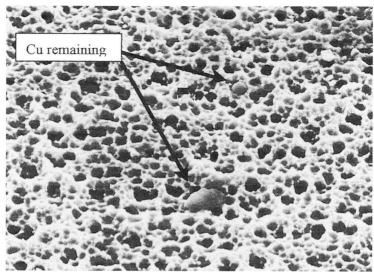

Figure 4. SEM micrograph of LCP surface shows copper nodules remaining after the copper etching process.

3.3 Vacuum-Metallized LCP Substrate

In addition to the LCP laminates, a direct vacuum metallization process has been developed to prepare LCP substrates. The process includes deposition of a thin conductor layer onto LCP film followed by electrolytic plating of copper to the desired thickness. Unlike LCP laminates, copper adhesion to LCP does not depend on surface roughness. Good adhesion of vacuum metallized substrate can be achieved without having a very rough surface, thus finer circuit pitches can be attained.

Vacuum-metallized polyimide films have been available for many years and have been used widely in flexible circuitry applications. Some of the technical challenges historically associated with the development of the process are: base film preparation, adhesion-promoting tie layer coating, initial copper layer deposition and copper adhesion. Direct metallization of LCP films has proven to be more challenging than metallizing polyimide films. Aside from all the technical challenges mentioned with polyimide film, some key issues that must be overcome are dimensional stability and web handling at relatively high web processing temperatures.

4   LCP FLEXIBLE CIRUIT PROCESS

Recent LCP circuit development offers significant improvements in manufacturing LCP flexible circuits and overcomes several limitations of circuit designs. The newly developed sputtering and chemical etching processes provide more freedom to manufacture additional varieties of circuit functional design features at low cost.

4.1 Process and Substrate Selection

4.1.1 Subtractive and Additive Processes

Either a subtractive or an additive process can be used to fabricate LCP flexible circuits. Each process offers its own advantages and disadvantages and there are similarities. Detailed fabrication process steps for polyimide flexible circuits are well known in the industry, and there are many similarities with the process steps used to produce LCP flexible circuits. Therefore, the discussion here will only focus on the differences. Process flow charts of both the additive and subtractive processes and the key steps are shown in Figure 5.

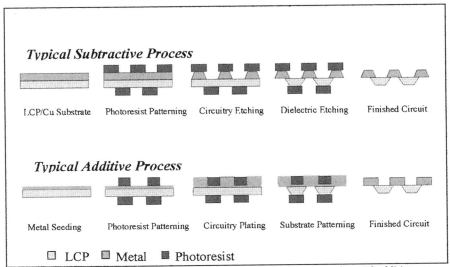

Figure 5. The major flexible circuit manufacturing steps with subtractive and additive process flows.

The subtractive process starts with either laminates or vacuum metallized substrates. Two types of copper foils may be selected for LCP laminates: electrodeposited (ED) copper foil or treated rolled/annealable (RA) foil. Typically, the copper foil contains a copper-oxide surface roughness enhancement, an encapsulation layer, a zinc-based thermal barrier layer, a chromium-based passivation or anti-tarnish layer and an organo-silane coupling agent. It may require an extra process step to remove these layers during the copper etching process.

The additive process normally starts with a vacuum-metallized substrate. Instead of copper etching, circuit traces are plated up, which gives a much straighter trace sidewall reflective to the photoresist profile and provides for a finer circuit pitch capability because there is less copper to etch. The choice for the circuit process between subtractive and additive depends on availability of the substrate type, circuit designs, processing capability and product requirements.

4.1.2 Laminates and Directly Metallized Substrates

Several test results have shown that circuits made from laminates suffered circuit trace delamination during temperature and humidity testing. Examination of the failed circuits showed a large undercut of traces that resulted from the need to remove excess metal from the roughened interface between the LCP and the copper (see Figure 4 and related discussion). In contrast, fine-pitch circuits that used the direct-metallized substrate made from an unroughened LCP surface performed well due to adequate adhesion of traces after vacuum metallization. The test results indicated that the metallized substrate is a necessary substrate for fine-pitch LCP flexible circuits.

Figure 6a presents a scanning electron micrograph of partial circuit traces of a LCP flexible circuit made from a laminate substrate. Figure 6b presents the same design LCP circuit made on a vacuum-metallized LCP substrate. The comparison clearly shows that LCP surface from sputtered circuit is much smoother than that from the laminate materials.

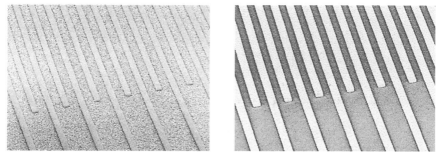

Figure 6. a) left: SEM micrograph of metal traces of LCP circuits made from a laminate substrate. b) right: SEM micrograph of metal traces of LCP circuits made from a sputtered substrate.

4.2 Sputtering Process

Metallization of LCP film is achieved by using a high-energy, vacuum sputtering process on a continuous, roll-to-roll basis. Copper can be sputtered directly onto LCP film, but a thin tie layer coating such as chromium or nickel may be used to enhance the bond strength between the copper and the LCP and to prevent copper migration at the interface.

Results from peel tests of the sputtered substrates range from 7-9 N/cm. The exact peel strength of the LCP substrate is difficult to measure because the LCP film fails cohesively. Thus, measurements actually indicate the cohesive strength of the LCP film. The true copper to LCP adhesion value should be greater than the LCP film cohesive strength. The detailed LCP circuit performance at high temperature and humidity, including dimensional stability and reliability for different end-use applications, will be discussed in the next section.

In summary, LCP film has been directly metallized using a vacuum sputtering process. Unlike laminated substrates, the sputtered LCP substrate yields excellent metal-to-LCP adhesion, avoids the trace (line) undercut problem, eliminates copper remaining during the etching process and provides effective circuit patterning resolution. Either subtractive or additive processing can be employed depending upon the requirement of the end-user.

4.3 Substrate Patterning

A key circuit process step is substrate patterning. Several methods, such as mechanical drilling, punching, plasma etching and laser ablation, have been used for producing vias and other features in LCP have been used. However, these methods are costly because they pattern one feature at a time, and involve expensive equipment that often needs to be set apart from the main production line and/or are inherently slow.

Chemical etching is a cost-effective method for patterning flex substrates and is more amenable for roll-to-roll processing comparing to other methods. Well known for producing features in polyimide circuits, chemical etching methods for producing via and through-holes in LCP has been complicated by the high resistance of LCP to chemicals. The discovery of new etchant for etching LCP is a breakthrough in recent circuit development [1].

The highly alkaline developing solution comprises an alkali metal salt and a solubilizer. A solution of an alkali metal salt alone may be used as an etchant for polyimide but it is ineffective for etching LCP in the absence of a solubilizer. Typically, the solubilizer is an amine compound. The efficiency of an amine in an etchant solution, according to the experiments, depends on its use with a relatively narrow range of concentrations of alkali metal salts including an alkali metal hydroxide.

The new etchant solutions allow continuous-process manufacturing LCP flexible circuits, which have fine features that were previously unattainable. Furthermore, complex circuit structures, such as unsupported cantilevered leads, through-holes and other shaped voids in films having angled sidewalls, can be produced using chemical etching. Typical chemically etched vias or through-holes have sidewall angles close to $45°$.

4.4 Advanced LCP Flex Circuits Summary

For the demonstration purposes, the recent LCP flex circuit advancements have been brought together in a nonfunctional flexible circuit design that illustrates a variety of different features (see Figure 7). These features include: chemically-etched vias down to 125 micron diameter in a 50 micron thick film, circuitry down to 30 microns pitch, a wire-bondable ball-grid array (BGA) circuit and protective coating patterning. Some of the specific features from this design can be viewed in more details in scanning electron micrographs (see Figures 8 and 9).

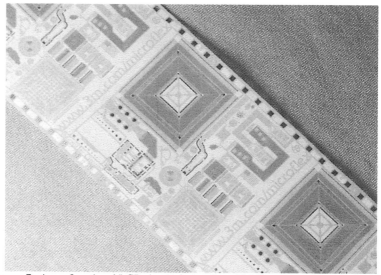

Figure 7. A nonfunctional LCP circuit design from 3M Company with a single metal layer that illustrates the variety of current manufacturing and design capability.

The circuit protective coating that has been applied to the circuit is actually a photoimagable cover coat but could also be a pre-patterned LCP coverlayer film laminated to the LCP circuit. Protective coatings are usually required by most applications to avoid corrosion, electromigration or other damage (e.g., physical) to circuit traces in order to protect against aggressive environmental or other end-use applications conditions. Protective coatings are also an important part of the circuit manufacturing process. Details of the process, coating types and test results will be discussed in the next section.

In summary, the key advancements described in this chapter represent a new complete process for LCP based on direct metallization, chemical etching and protective coating, which makes LCP flexible circuits available and within the reach of many demanding applications.

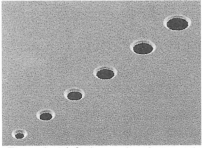

Figure 8. a) left: A scanning electron micrograph with the typical chemically etched through-holes at different sizes. b) right: An array of chemically etched vias

Figure 9. Unsupported cantilevered leads from a partial LCP circuit

5 RECENT TEST DATA AND EARLY LCP CIRCUITS APPLICATIONS

Test data from a series of development studies has demonstrated that LCP is a practical flex circuit substrate with excellent dimensional stability, material compatibility and manufacturability when processed beyond circuitization and combined with other materials such as LCP coverlayers or soldermasks in final product constructions. Specific activities in this section include metal and soldermask adhesion tests, an entire battery of reliability tests and determination of the circuit interactions with common assembly processes of die attachment, wire bonding, encapsulation and solder reflow.

5.1 Dimensional Stability Studies

As discussed in section 2.1, LCP is different from polyimide in its bi-directional structure (molecular chains are oriented in the machine direction

"MD" and perpendicular to the transverse direction "TD") and its slightly lower transition temperature. Because LCP is a thermoplastic, roll-to-roll processing presents some unique challenges. Tension and temperature can potentially affect the circuit dimensions, especially with the higher temperatures used in LCP coverlayer film laminations of LCP circuits (> 280°C). The dimensional work was divided into three separate efforts: (1) circuit manufacturing, (2) LCP coverlayer lamination and (3) temperature-time effects on the completed circuits that can be related to different reliability tests and assembly process conditions.

Critical feature dimensions of a single test circuit design were measured at multiple MD and TD locations from the web after several circuit and lamination runs that involved different LCP raw material lots. Differences between the critical LCP circuit dimensions and the glass phototool were analyzed in order to extract raw material and process effects and to compare with the existing polyimide circuit specification of +/- 0.2% [18]. After exposure to the different process conditions, the circuits were always brought back to the manufacturing environment (25°C, 40% RH) before measuring the dimensions.

5.2 Assembly Process Compatibility Studies

The assembly process compatibility studies concentrated on solder reflow and wire bonding of LCP circuits. Worst-case temperatures of 280°C and 260°C for no-Pb solders were used in liquid immersion and in-line reflow experiments, respectively. For the wire bonding study, the bond quality was determined and the process window of LCP circuits was compared with that of polyimide.

For metal adhesion tests, trace-shaped test patterns were etched on metallized (sputtered and copper-plated) LCP films from different lots. Peel adhesion values were recorded before and after exposure of specially prepared coupons (3 coupons per film type and treatment) to molten solder for 5 seconds at 280°C [19] as a function of film orientation. For the dimensional experiments, test circuits were used and molten solder was replaced by an inert perfluoropolyether (Galden® D-02TS, Ausimont USA, Thorofare, NJ). The temperature was also 280°C, but the exposure was 10 seconds.

For in-line reflow experiments (3X reflow is specified for MRT reliability testing [20]), convection oven (Vitronics-R Unitherm SMR-400) profiles were adjusted to create separate 240°C and 260°C peak temperature profiles. Both had 90 seconds above the liquidus (183°C and 217°C, respectively).

For the wire bonding experiments, a Kulicke & Soffa 1488-plus was used to form both ball and stitch bonds on LCP test circuits (513 total wires bonded on 23 total circuit coupons with two different designs). Starting with parameters used for bonding polyimide flex circuits, the force and power parameters were changed in order to bring the bonding tool imprint and bond strength up to levels that resulted in good bonds (average pull strength greater than about 10g of force), then the time was adjusted downward to 40msec. Finally the temperature variable was changed over the 100°C to 175°C range and average pull strength and failure mode were analyzed. Pull tests were accomplished with the hook located in the middle of the wire bonds.

5.3 Materials Compatibility Studies

5.3.1 Sample Preparation Matrix for LCP Circuits with Soldermasks and LCP Coverlayer Films

The material compatibility study investigated the interactions of soldermasks and LCP coverlayers with LCP circuits before and after reliability testing. The adhesion of soldermasks (blanket-coated, also flood-exposed for photoimaged systems) with LCP films was measured as a function of soldermask processing condition (e.g., precondition, application, expose, dry, cure). Five soldermasks, representative of a wide variety of different product applications, were included from the major polymer chemistry families, including epoxy, epoxy-acrylate and urethane.

LCP coverlayer films were laminated to LCP circuits at different temperatures (above the softening point of LCP) and pressures, after the coverlayers were first patterned using soft tooling. Both Type 1 and Type 2 LCP films, which differ in their thermal properties, have been used both as raw materials for circuits and for coverlayer films.

5.3.2 Battery of Reliability Testing

Both LCP circuit constructions with soldermasks and with LCP coverlayers were subjected to a similar battery of reliability tests, but the tests for the latter were more severe (e.g., Level 1 "L1-MRT" versus Level 3 "L3-MRT" [20]) to demonstrate their greater expected moisture resistance. For the soldermask experiments, flood-exposed coatings on LCP films were evaluated visually and with peel testing both before and after the following battery of reliability test conditions:

1) Moisture Resistance Tests: Pressure Cooker Test, 168 hours [21], Highly Accelerated Stress Test at 130°C (Unbiased HAST, [22]), Moisture Resistance Test, Level 3 (MRT) [20]

2) Moisture-Bias Tests: HAST [23] at 130°C and 5 volts for 96 hours (exception: no peel test and no bare films, but instead an interdigitated circuit with adjacent traces biased oppositely was used; the equipment would monitor any leakage current and would automatically stop for greater than 5 microamps)

3) Thermal Shock (TS) Tests: Liquid-to-Liquid, -55°C to 125°C [24] and Air-to-Air, -65°C to 150°C [25] for 1000 cycles

4) Liquid Immersion Tests (e.g., solder), 280°C [19, 26]

5.3.3 Soldermask Peel Testing

For peel tests, adhesion was measured either with 180° (see [27]) or T-peel tests of 12.5mm-wide specimens cut from larger rectangular pieces of LCP that had been blanket coated with soldermask. In the 180° test, specimens were mounted with the soldermask side glued with strong adhesive onto rigid boards. After separating the LCP from the covercoat manually in a short section, the LCP was pulled away from the specimen / board fixtured at a 180° orientation from the crosshead direction (Instron, model number 5567; for test set up, see [27]). The failure mode was either adhesive at the LCP / soldermask interface or strictly in the LCP, and the average or peak force, respectively, were recorded. LCP failures often exhibited a mixed failure mode with some LCP attached on top of the soldermask layer that was accompanied by stretching, thinning and tearing of the remaining LCP in tension, thus the peak force recorded in that case was considered to be a lower limit of the true soldermask-to-LCP adhesive strength.

In the T-peel test, strong polyester adhesive tape was attached to the soldermask side after the specimen had been pre-scribed in the soldermask layer only (not in the substrate [28]), and the crosshead of the Instron gradually pulled the tape off of the specimen as it was held in a fixture at the same 180° angle orientation as in the 180° test. As the peel proceeded, the angle of separation ranged from 90° (typical, especially found when the peel line reached the scribed lines) up to 180°. As the specification [28] allows, soldermask adhesion with LCP measurements in this report were semi-quantitative and based on a 6-point scale, which at the extremes had 5, denoting no soldermask removal on the tape, and 0, denoting over 65% of soldermask pieces removed in the grid pattern (actually "5B" and "0B", [28]). Because smaller pieces of soldermask could be more easily removed on account of the scribing with this T-peel test, it provided a more sensitive means of determining relative adhesion for these specimens, when more direct measure of the adhesion force was impossible to measure with the 180° test because of LCP failures. As compared to the 180° test, the T-peel test made LCP failures less likely and was more sensitive to smaller areas on the specimen having poorer adhesion than other areas.

5.4 LCP Circuits Dimensional Stability Results

Finished test circuits from two runs (see Table 2) showed a slight shrinkage (negative bias) in the transverse direction (TD) but little movement in machine direction (MD). Thus, different global compensations in the circuit designs may be warranted for MD and TD, which were unnecessary for polyimide. However, the variability in the measurements (3 sigma values) was low and easily met the +/- 0.2% specification established for polyimide [18].

Table 2. Critical Dimensions Statistics of Two Circuit Runs with the Test Circuit Design

LCP Raw Material Lot Number	003	004
Sputter Run Number	10	12
Circuitization Run Number	566	567
MD: 3 Sigma	0.058%	0.059%
MD: Bias	+ 0.008%	+ 0.003%
TD: 3 Sigma	0.061%	0.069%
TD: Bias	- 0.048%	- 0.033%

After exposure to all the different reliability tests, critical dimension changes on these finished test circuits were negligible and typically an order of magnitude less than the +/- 0.2% specification [18]. This was observed whether the tests involved long times at lower temperature or short times at higher temperatures (see first 5 and last 3 tests, respectively, in Table 3). Similarly, assembly processes like adhesive and overmold cure and wire bond are expected to cause negligible LCP circuit dimensional changes. Die adhesive and overmold cure and wire bond conditions are like the first 5 tests in temperature (125-150°C) but involve much shorter time (<0.5 to 4 hours), so there would be less dimensional change than observed for these tests. Analogously, molten solder processes like solder ball attach or component reflow, which are like the latter three tests, would exhibit similar dimensional changes.

Table 3. Dimensional Changes of Test Circuits after Exposures to Different Reliability Test Conditions

Maximum Temp. (°C)	Time at Maximum Temp. (hours)	Reliability Test Description	Bias - MD (percent)	Bias - MD (percent)
125	25	Liquid TS, -55 to 125°C [9]	-0.015	+0.006
125	57.5	``	-0.008	+0.007
130	96	HAST, 130°C, 85% RH, 96 hr [7]	-0.009	+0.023
150	120	High Temp. Storage, 150°C	-0.009	+0.014
150	221	``	-0.017	+0.012
260	~0.003	3X Reflow at 260°C	-0.011	+0.032
260	~0.003	130°C, 85% RH 96 hrs plus 3X reflow at 260°C	-0.020	+0.055
280	~0.003	10 sec inert liquid immersion [5,11]	-0.043	+0.062

5.5 Wire Bonding Results

Wire bonding of LCP circuits from two different circuit runs and designs was successful as evidenced by large average pull tests. Failure modes overwhelmingly arose from wire breakage and not stitch or ball bond lifting (see Figure 10). Bonding on LCP circuits appeared to have a similarly sized process window relative to polyimide flex with the initial circuit designs, but later results using other designs, still under analysis, indicate that bonding is also influenced by bond finger size and wire diameter. With the initial circuits, the best results were achieved at 150-175°C temperature settings. A representative cross-section is included in Figure 11.

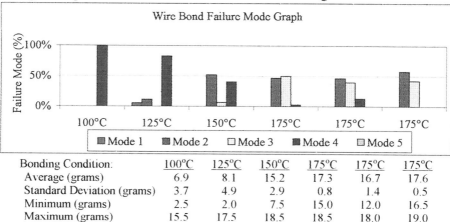

Bonding Condition:	100°C	125°C	150°C	175°C	175°C	175°C
Average (grams)	6.9	8.1	15.2	17.3	16.7	17.6
Standard Deviation (grams)	3.7	4.9	2.9	0.8	1.4	0.5
Minimum (grams)	2.5	2.0	7.5	15.0	12.0	16.5
Maximum (grams)	15.5	17.5	18.5	18.5	18.0	19.0

Figure 10. Wire Bonding Pull Test Statistics: Averages versus Process Temperature and Failure Mode Distribution for LCP circuits. Modes 1 and 5 were ball and stitch bond breaks (undesirable) and Modes 2, 3 and 4 were wire breaks (desired) above ball, at hook and at stitch.

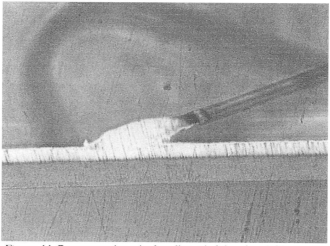

Figure 11. Representative wire bonding stitch bond on LCP circuits

5.6 Molten Solder Compatibility Results

Copper trace adhesion after solder float was equivalent to before solder float within error (ratio average 101%, see Table 4). Thus, flexible circuits made with LCP can withstand soldering processes and are suitable for applications requiring soldering.

Table 4. Metal Trace Adhesion Comparison Before and After Exposure to Molten Solder

Sample Number and Orientation of Trace	Substrate Lot Number	Copper Thickness (μm)	Percent Peel Strength Retained after Solder Exposure *
11A, MD	002	17	100%
11A, TD	002	16	110%
12A, MD	002	18	105%
12A, TD	002	18	102%
13A, MD	001	19	99%
13A, TD	001	19	112%
11B, MD	002	17	96%
11B, TD	002	16	104%
12B, MD	002	18	90%
12B, TD	002	19	96%
13B, MD	001	18	96%
13B, TD	001	20	105%

* Percent Peel Strength Retained = (Average Peel Strength after Solder Float Treatment / Average Peel Strength without Solder Float Treatment) x 100%

5.7 Soldermask Compatibility Results

5.7.1 Visual Inspection & 180-degree Peels

In the initial screening for the soldermask process condition and reliability test factors in the experiment, condition 1 (see Figure 12, data sets with small squares) and pressure cooker (PCT) and HAST gave the weakest peels. Condition 1 always gave an adhesive failure mode, with some only 1-10% of their as-cured values after PCT or HAST. This was in agreement with the less-sensitive visual inspection, which showed instances of delamination only with these factors and with soldermask 1 and 3 only.

Condition 2 (data sets with large squares) was considerably better, giving consistent high peel strengths (in the 1 to 3 units range) and almost exclusively LCP failures even after the most severe reliability tests (PCT and HAST). This indicated a strong adhesive bond between soldermask and LCP that was not weakened during testing.

Compared to polyimide, the soldermask interface strength with LCP (measured as peak force) using the best soldermask process condition met or exceeded that with polyimide (for example, soldermask 3 adhesion to polyimide was 0.2 units, as compared to 2.3 units on LCP). During the tests, the LCP substrate tore more frequently than polyimide, but this was easily explained by differences in LCP and polyimide mechanical properties. It should be noted that when condition 1 was used, similar reductions in adhesion strength of soldermask to polyimide was observed after PCT testing, as was previously described for LCP.

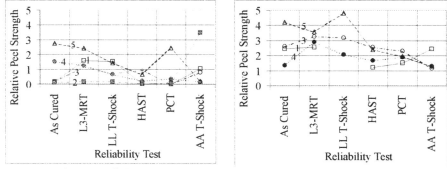

Figure 12. Relative 180° Peel Test Values for 5 Different Soldermasks (epoxy –1, epoxy-acrylate-2,3,4, urethane-5) with Different Processing Conditions 1 (left) and 2 (right)

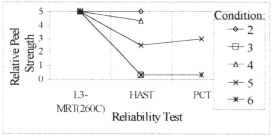

Figure 13. T-Peel Semi-Quantitative Soldermask-to-LCP Adhesion Values (0=poor adhesion, 5=good adhesion [13]) for Soldermask 1 using Various Process Conditions

5.7.2 T-Peel Test Results

After the screening experiments, additional soldermask conditions 3 to 6 were investigated with the T-peel method under worst-case reliability test conditions (Level 3 "L3-MRT" with 260°C reflows preconditioning and HAST or PCT). Figure 13 shows data for soldermask 1, but all soldermasks gave similar results. Strong adhesion was observed after L3-MRT with no differentiation between conditions 2 to 6, but clear differences between conditions arose when the samples were exposed to severe moisture reliability tests (HAST or PCT). The preferred process condition order for soldermask 1 was clearly: 2>4>5>>3~6>>1 and was consistent with 180° peel testing. The best process conditions for the other 4 soldermasks was similarly condition 2, which also gave relative T-peel strength values of 5.0. For the 180° tests, no real discernment in soldermask adhesion level to LCP could be made between specimens for conditions 2 to 6 (data not shown), because all exhibited LCP failures, which only indicated that they exhibited a high adhesion level.

5.8 LCP Coverlayer Film Laminated LCP Circuits

Examples of typical cohesively bonded structures observed after lamination are shown in Figure 14. LCP test circuits laminated using six different lamination conditions (1-6) passed the preliminary biased HAST test (see Table 5). After limiting conditions to two alternatives (7-8), these circuits passed Level 1 MRT and all the other reliability tests, which indicated adequate LCP-to-LCP bonding.

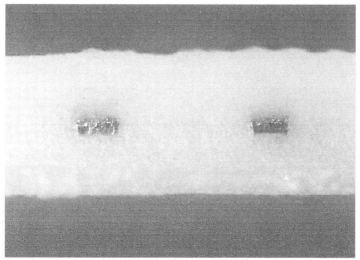

Figure 14. Cross-section view of LCP coverlayer film (top) laminated to LCP circuits (bottom)

The dimensional stability of LCP-circuits after lamination of LCP coverlayer films is anticipated to meet existing polyimide circuit specifications (< +/- 0.2%, see Table 6, case A). The 3-sigma values were higher (0.09-0.14%) than those measured after circuitization (0.05-0.07%, see Table 1), but this was primarily because the lamination temperature was higher than any circuitization process step, in fact even higher than molten solder immersion (280°C). Similarly the bias in MD and TD (around 0.1%, not shown in Table 6) was also higher than observed with any lower temperature exposure (0.05% and 0.06% were maximums in the absolute value of the bias measured after circuitization and reliability testing, respectively, see Tables 2 and 3). Compensation of the design due to bias is planned in order to make up for LCP film dimensional changes caused by lamination near its transition temperature.

Table 5. LCP Coverlayer Film Laminated to LCP Circuits. All Parts Passed all Reliability Tests in a Wide Process Window.

Lamin-ation Condition	Bias HAST [8]	Lamin-ation Condition	Un-biased HAST [7]	Pressure Cooker Test [6]	Level 1 MRT [5]	Liquid-to-Liquid Thermal Shock [9]	Air-to-Air Thermal Shock [10]
1	3/3	7	6/6	6/6	10/10	6/6	6/6
2	3/3	8	2/2	4/4	4/4	4/4	4/4
3	3/3						
4	3/3						
5	3/3						
6	3/3						

Comparing case A with case B (see Table 6) indicates that design constraints can be successful to lower the variability in dimensions to acceptable levels. The variability in circuit dimensions after coverlayer lamination was in general a function of both the process (lamination) condition and the design. The variability due to design arises because of different critical feature locations in the layout and was found to be sensitive to locations near features such as large areas of etched dielectric. The variability for case A was almost exclusively due to process and not design factors, but this was exactly the opposite for case B (3 Sigma >> 0.2%). In fact, the variability due to the process for case B was about equal to the total variability of case A. Thus, applications requiring the high degree of moisture protection and chemical inertness afforded by strong cohesive LCP-to-LCP bonding that the LCP coverlayer-laminated LCP circuits achieve may require some design constraints in addition to optimizing lamination process conditions to maximize dimensional stability of the final circuit construction.

Table 6. Variability of Critical Dimensions for Different Lamination Conditions

Coverlay Lamination Condition Number	1	2	3
Material Lot Number	003	003	005
Sputter Run Number	10	10	9
Circuit Run Number	566	566	565
Number of Lamination Runs at this Condition	15	15	30
Case A: 3 Sigma (constrained design)	0.089%	0.134%	0.094%
Case B: 3 Sigma (unconstrained design)	0.333%	0.314%	0.310%

5.9 Early Applications Interest for LCP Circuits

Integration studies with multiple materials are on-going for many product constructions to meet a variety of different application needs. Results from the integration of soldermasks (see Table 7) predict that any soldermask material required by a specific application probably will prove to be compatible with LCP circuits. This is because acceptable process conditions for blanket coatings on LCP films have been defined for all three chemical categories of soldermask tested. Hard disk drives (HDD) represent one such application area that use a photoimageable soldermask and might find LCP circuits a suitable alternative to polyimide. This is because both the low CHE and CTE matched to copper may allow for flatness over a greater range of operating conditions, and LCP mechanical properties make for improved read/write head flight characteristics versus polyimide.

Table 7. Summary of LCP-Coverlayer and Soldermask Reliability Results on LCP Circuits

Battery of Reliability Tests			Coverlayer (LCP)	Soldermask *
Test	Condition	Time		
Pressure Cooker Test (PCT) [6]	121°C, 100%RH	168 hours	PASS	PASS
Thermal Shock: Liquid-to-Liquid [9]	-55 – 125°C	500 cycles	PASS	PASS
		1000 cycles	PASS	PASS
HAST [7,8]	130°C, 85%RH, 5 & 0 volts	96 hours	PASS	PASS
Thermal Shock: Air-to-Air [10]	-65 – 150°C	500 cycles	PASS	PASS
		1000 cycles	PASS	PASS
Level 1 MRT [4]	85°C, 85%RH + 3X Reflow 260°C	168 hours	PASS	Not tried
Level 3 MRT [4]	30 °C, 60%RH + 3X Reflow 260°C	192 hours	Not tried	PASS
Liquid Immersion (e.g., solder) [5,11]	280°C	5 – 10 sec.	Not tried**	PASS**

* 5 soldermasks evaluated on LCP film (including epoxy, epoxy-acrylate, and urethane polymers)
** Raw Circuits – PASS (metal adhesion and dimensional stability)

The positive moisture resistance results obtained thus far with soldermasks and LCP coverlayers, as summarized in Table 7, may indicate a fit for LCP flex circuits in some emerging applications that demand high moisture resistance. Many applications that require solder reflow (e.g., ball grid arrays – BGAs – and flip chips) and bias voltage (3-20 volts) / high humidity reliability testing (85C/85%RH or more severe, HAST [22]) should examine LCP in addition to polyimide, which is traditionally used, because the high-humidity moisture tests like HAST become more difficult to pass with the migration of these applications to finer pitch (30µm fine pitch flex circuits in development). Biomedical devices, such as hearing aids, special detectors or other devices, such as salinity sensors, and inkjet printers represent other applications that require flex circuits to be resistant to corrosive environments (e.g., body fluids and inks). Theoretically, stresses on interfaces during aqueous tests are expected to be much less with LCP than with polyimide flex circuits because the CHE and moisture uptake were more than one-half to one full order of magnitude less, respectively, for LCP than for polyimide (see Table 1). Indeed, promising results have recently been obtained in tests more severe than biased-HAST tests, where moisture is applied by immersion of coverlayer-laminated circuits into corrosive liquids (instead of high humidity environments as in HAST) and a bias voltage is applied to try to cause electromigration failures.

This same moisture inertness gives rise to another benefit of LCP over polyimide for high frequency electrical applications. Polyimide appears to be useful at frequencies up to about 10 Gb/sec for circuit lengths less than 12 mm for optoelectronic applications. However, use of polyimide as a flex circuit substrate material at higher frequencies and with longer circuit lengths will likely to be limited by humidity. Both the dielectric constant and dissipation (loss) factor for polyimide have been reported to increase with humidity (3.5 to 3.7 and 0.009 to 0.018, respectively, from 0 to 90% RH [29]), while those for LCP are humidity-independent (3.0 and 0.003 at both 0 and 90% RH [29]). Indeed, the dielectric stability and loss of LCP at higher frequencies (e.g., 45 GHz) are more like those of polytetrafluoroethylene-based (PTFE) materials than polyimide [14,30], so LCP circuits may be valuable for microwave circuits (e.g., radar applications). In comparison to PTFE-based materials, however, LCP permits simpler processing with potential cost savings, such as patterning through chemical etching in addition to laser ablation. For high frequency applications, advanced LCP flex circuits offer fine-feature capability (fine lines and spaces and dielectric patterning) of polyimide flex circuits with a negligible moisture effect, which will permit high frequency circuitry applications with reduced propagation delays and cross talk between signal lines.

In summary, the unique properties and features offered by the advanced LCP circuits should extend applications and design opportunities beyond the current capabilities of polyimide circuits, conventionally produced LCP circuits (made from film lamination) and circuits produced from substrates with coarser or more expensively and difficultly produced features (e.g., PTFE). The advanced LCP flexible circuits will quickly find a wide spectrum of new applications as end-users discover their unique characteristics in electronic device design.

6   LCP CIRCUITS AND FOLDABLE FLEX MULTICHIP PACKAGING

Although attractive to a variety of different application areas (see section 5.9 in this chapter), LCP substrates have not been used in any of the emerging foldable-flex, multichip-packaging applications thus far to the authors' knowledge. However, recent technical advances with LCP flex circuits warrant their consideration. Many existing foldable polyimide flex designs make use of 1ML (one metal layer) technology, which has been reviewed for the LCP substrate in the first five sections of this chapter, but most require 2ML technology (see Figure 15). This section provides a

from soldermasks and LCP coverlayer films (see Table 7) to others, which have adequate adhesion under moisture exposure.

4) Other circuit designs passing severe bias HAST tests: These results provide complementary assurance for LCP to previously reported HAST successes using constructions based on 1ML films coated with soldermasks (see Table 7).

5) Assembly of 12-mm BGA packages (see Figure 15) with supporting dimensional stability and reliability (Level 3 JEDEC moisture resistance testing [20]) data: These data on more complex constructions are consistent with previous tests on simpler raw circuit and film constructions (see Tables 3 and 7, respectively).

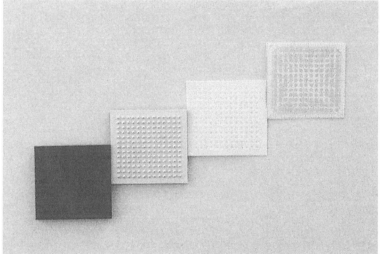

Figure 16. 1ML 12mm BGA circuits made from LCP substrate with 50µm minimum pitch, that is being used as a test vehicle to illustrate LCP circuits compatibility with common assembly materials and processes. The photo includes from top right to bottom left: metal circuitry side, LCP dielectric side, solder ball side of package, encapsulated side of package.

In view of the completeness of these results, there are few technical gaps in the 1ML LCP flex circuits technology relating to foldable flex, multichip packaging. Gaps are believed to be primarily related to the specific construction and materials that are normally chosen by the end user and/or assembly manufacturer during joint studies with the flex circuit supplier. In the reliability assurance area, an essential demonstration of the moisture resistance and bias testing of the entire specific package construction, which includes specific materials (folded flex adhesives, encapsulants, overmolds, etc., see also Figure 15), is needed. Also, specific foldable-flex designs need to be run through the assembly manufacturer's process equipment to verify test results. Differences in the manufacturing set-ups from those used for polyimide films are expected because LCP properties (chiefly mechanical ones) differ somewhat with polyimide.

Folding of LCP circuits is one technical gap in the assembly process. However, LCP is expected to perform at least equivalently to polyimide based on cycling tests of raw films (see Table 8) [30], but these tests need to be verified with LCP circuits according to the requirements of specific folded flex applications. Besides the resistance of the dielectric material to fracture, the test results are dependent on other factors including the adhesion of the traces and any soldermask or coverlayer films attached to the dielectric in the specific design.

Table 8. Modulus and fracture resistance to flexing of different flexible circuit films

	LCP Type 1	Polyimide 1	Polyimide 2
Modulus (kgf/mm2)	700	350	900
Flex Resistance, MIT (cycles), r=2.0mm	5×10^6	2×10^6	1×10^6
Flex Resistance, MIT (cycles), r=0.25mm	20×10^3	10×10^3	5×10^3

6.2 Two-Metal Layer (2ML) LCP Flex Circuits

LCP flex circuits can be bent, folded, or shaped to interconnect in multiple planes or conform to specific package sizes. As discussed in the previous sections, 1ML LCP flex circuits provide many excellent features and have some additional benefits over polyimide circuits. Besides the simple, single-metal designed configurations, LCP flex circuits can be designed to more complex three-dimensional conductive paths, two-metal (2ML) flex circuit assemblies.

The fabrication of 2ML LCP circuits is more complicated than processing 1ML circuits. 2ML process involves additional steps such as second side metallization and some of the processes are much more challenging. Via making and conductive connections through the vias are very important to product reliability. Dimension stability of circuits is very crucial to the second side process. It becomes more critical for selecting proper process flow such as subtractive versus additive, laminates versus metallization and chemical etch versus punch or laser drilling.

2ML polyimide flex circuits are manufactured routinely as standard products. The expertise from raw materials to circuit fabrication and value-added assembly is being applied to develop 2ML LCP flex circuits at 3M Company. As an example, Figure 17 shows a via cross-section view from both 1ML and 2ML LCP flex circuits. The copper layers on both sides of the dielectric film are plated continuously and the thickness can be achieved at any specification. The chemical etching method may be applied to form vias as well as complex and fine features while avoiding costly punching and laser drilling. Figure 18 shows an example of different via sizes formed by chemical etching method.

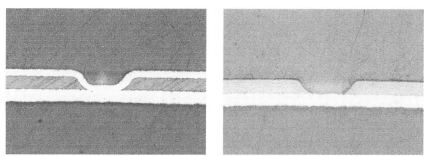

Figure 17. LCP flex circuit vias cross-section view of 2ML (left) and 1ML (right)

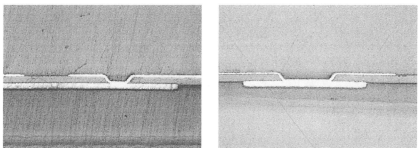

Figure 18. 2ML LCP flex circuits cross-section view of different via sizes

6.3 Summary

LCP flex circuit technology and applications have been reviewed. Significant recent technology advancements include vacuum sputtering for fine lines and spaces and chemical etching for low-cost dielectric patterning. Recent test data on dimensional stability and reliability indicate small dimensional changes with moisture and compatibility with many assembly processes and materials. Based on this data, LCP flex circuits have attracted interests in a wide variety of product areas including demanding high frequency designs and applications requiring high moisture and chemical resistance.

The technical readiness for LCP flex circuits for the foldable flex, thinned-silicon, multichip packaging applications has been reviewed. 1ML LCP technology is ready now for investigations of specific constructions. 2ML LCP circuit development is not as ready but has made much recent progress with second-side metal sputtering, because of similarities with vias and second-side plating processing and equipment from the existing 2ML polyimide technology infrastructure. Besides the laser ablation method of drilling vias practiced primarily with polyimide, 2ML LCP circuit processing has the potential of making vias with a lower cost by the chemical etching of fine features in the dielectric. LCP should become a complementary flex circuit offering to polyimide for many emerging application areas, including folded flex multichip packaging, as its capabilities for advanced designs becomes better known to end-users.

7   REFERENCES

1. R.Yang, "Liquid Crystal Polymers for Flexible Circuits," Advanced Packaging, March 2002. (portions of this text were first published in the March 2002 issue of Advanced Packaging magazine, copyright 2002 PennWell Corp.)

2. T. Hayden, "New Liquid Crystal Polymer (LP) Flex Circuits to Meet Demanding Reliability and End-Use Applications Requirements," Proc. International Conference on Advanced Packaging Systems (ICAPS), Reno, Nevada, March 2002, pp. 116-122.

3. F. Reinitzer, Monatsh. 9, p421, 1888

4. "Liquid crystalline polymers to mining applications", Encyclopedia of polymer science and engineering, 2nd ed., Vol. 9, editorial board, Heman Mark, et al.

5. L. Chapoy, Recent Advances in Liquid Crystal Polymers, Elsevier, 1985

6. A. Donald and A. Windle, Liquid Crystalline Polymers, Cambridge University Press, 1992.

7. M. Gordon, ed., Liquid Crystal Polymers I, Vol. 59, and Liquid Crystal Polymers II/III, Vol. 60/61 of Advances in Polymer Science Series, Springer-Verlag, New York, 1984.

8. J. Appl. Polym. Sci. Appl. Polym Symp. 41, 1985.

9. R. Lusignea, K. Blizard, R. Haghighat, Extrusion of High Temperature Thermoplastic Liquid Crystalline Polymer Microcomposites, Ultralloy `90 Symposium, Shotland Business Research, p151, 1990.

10. R. Lusignea, Polymer Engineering and Science, Dec. 1999, Vol. 39, 12

11. U.S. Pat. No. 5534209

12. U.S. Pat. No. 6274242 B1

13. C. Bergstrom, Polymers and Other Advanced Materials: Emerging Technologies and Business Opportunities" Plenum Press, New York, p777, 1995

14. G.Zou, H. Gronqvist, P. Starski, and J. Liu, "High Frequency Characteristics of Liquid Crystal Polymer for System in a Package Application", Proc. 8th International Symposium on Advanced Packaging Materials, pp 337, 2002

15. J. Economy and K. Goranov, "Thermotropic liquid crystalline polymers for high performance applications," Advances in Polymer Science, Vol. 117, p 221, 1994.

16. M. Lawrence and Y. Tanaka, "Copper adhesion to liquid crystal polymer and other film-based circuit substrate", p354, Internaltional Symposium on Microelectronics, 1999.

17. P. Nevrekar, "Liquid Crystalline Polymers: The next Generation of High Performance Flex Circuit Materials", CircuiTree, April 2002

18. 3M Company Publication Number 80-6103-6334-5, "3M Company Specifications and Design Guidelines: Microflex Circuits for Interconnect Solutions", Section VII, 1997.

19. IPC-TM-650 Method 2.4.9, Revision D, 1999.

20. IPC/JEDEC J-STD-020A, "Moisture Reflow Sensitivity Classification for Non-hermetic Solid State Surface Mount Devices," Joint Industry Standard, April 1999.

21. JESD22-A102-C, "Accelerated Moisture Resistance – Unbiased Autoclave," JEDEC Standard, December 2000.

22. JESD22-A118, "Accelerated Moisture Resistance – Unbiased HAST," JEDEC Standard, December 2000.

23. JESD22-A110-B, "Highly Accelerated Temperature and Humidity Test (HAST)," EIA/JEDEC Standard, February 1999.

24. JESD22-A106A, "Thermal Shock," JEDEC Standard, April 1995.

25. JESD22-A104-B, "Thermal Cycling," JEDEC Standard, July 2000.

26. ASTM D2732-96, "Standard Test Method for Unrestrained Linear Thermal Shrinkage of Plastic Film and Sheeting," approved September 10, 1996.

27. ASTM D903-98, "Standard Test Method for Peel or Stripping Strength of Adhesive Bonds," approved April 10, 1998.

28. ASTM D3359-97, "Standard Test Methods for Measuring Adhesion by Tape Test," approved November 10, 1997.

29. H. Inoue and S. Fukutake, "Liquid Crystal Polymer Film with High Heat Resistance and High Dimensional Stability," SMTA Pan Pacific Microelectronics Symposium, February, 2001.

30. Gore product information data sheet of BIAC LCP flex dielectric

Chapter 12

FLEX AND THE INTERCONNECTED MESH POWER SYSTEM (IMPS)

Leonard W. Schaper
University of Arkansas

12.1 INTRODUCTION

Flex is essentially a double-sided wiring technology. It is certainly possible to multilayer flex, but this adds considerably to the cost and deviates from the theme of this book, which is foldable flex. Once multilayered, flex is no longer foldable. So this chapter will concentrate on what can be done, electrically, with one sheet of double-sided flex, with fine lines and laser-drilled vias. For discussion purposes, assume a fabrication capability for 50 µm line and space, and 50 µm laser drilled vias in 150 µm catch pads.

Traditionally, double-sided flex has been considered as a low performance interconnect medium, since it does not provide the electrical functionality of four metal layers, namely power plane, ground plane, X signal, and Y signal. On flex, power distribution is done on traces, albeit wide traces, so the low-impedance structure of closely spaced planes is not available. Moreover, lacking a ground reference, either for microstrip or stripline topologies, signal conductors are not constant impedance transmission lines. Depending on their proximity to an AC current return line, they could have a variety of impedances, and could certainly exhibit substantial crosstalk. However, it is possible to extract a high performance interconnect medium from two physical layers of metal.

12.2 IMPS

The Interconnected Mesh Power System (IMPS) is a systematic topology that allows low inductance planar power and ground distribution; and dense, controlled-impedance, low crosstalk signal transmission in as few as two physical wiring layers. It utilizes the production methods of fine line lithography and batch via fabrication characteristic of MCM-D and some MCM-L to create a structure not economically possible using standard printed wiring board methods. Though IMPS was originally implemented in MCM-D substrates on silicon or glass, it has also been implemented on flex, as will be described.

12.2.1 The IMPS Structure

The derivation of the power distribution structure of IMPS is shown in Figure 1. Consider a familiar meshed plane, used in many MCMs for power or ground. Think of this construction not as a plane with holes, however, but as a set of X and Y conductors. In Figure 1a, the X conductors and Y conductors have been placed on two separate metal layers, and at each crossover, a via has been provided to retain the planar characteristics. (Vias typically have low resistance and inductance compared with lines.) This "interconnected mesh" plane is reasonably electrically equivalent to, but topologically different from, the conventional mesh plane.

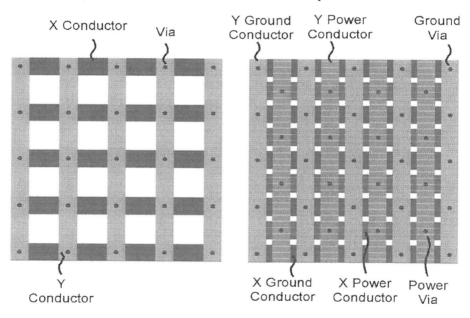

The mesh of Figure 1a includes enough space between conductors to insert interdigitated conductors of opposite polarity (power or ground) with their own set of interconnecting vias, on the same two physical metal layers (Figure 1b). The resulting structure forms a complete power distribution system.

In the simplest IMPS implementation, the minimum width power distribution mesh conductors are placed at twice minimum design rule pitch, leaving signal wiring tracks between each P-G pair. If a track is not used for signal routing, the power carrying conductors are widened into the empty space, for the maximum metal coverage, as shown below in Figure 2. The resulting construction provides a constant impedance signal transmission environment with low crosstalk, as will be discussed.

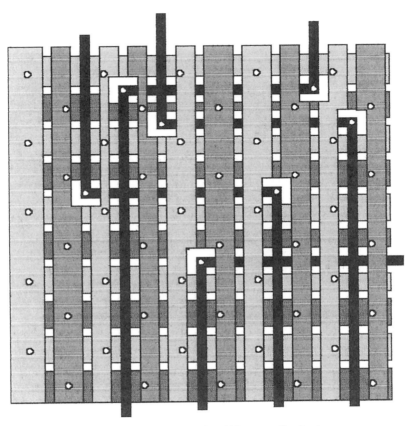

Figure 2. IMPS using variable width power distribution.

IMPS derivatives can also be developed for pre-defined grids of vias, or other structures, using the same underlying principles. IMPS has been extended to multiple metal layers, and to the distribution of several voltages. IMPS design rules have been developed for many fabrication methods and sets of materials. IMPS is a general topology, not a specific set of rules only applicable to, for example, MCM-D.

12.2.2 IMPS Benefits

In the case of using IMPS to reduce a multilayer structure to a two-layer structure, IMPS has several benefits. The obvious benefit is the reduction in manufacturing process steps and the associated direct cost. The typical 4-layer structure implemented in MCM-D uses 8 masks; IMPS needs only 4 masks. There is, of course, a yield improvement also resulting from fewer process steps. The effect on the factory is to halve thruput time and double manufacturing capacity.

For foldable flex applications, the signal traces with the interleaved power and ground circuits still have the physical result of only one flex substrate with double sided circuits. This construction is still flexible and foldable even though it contains the power and ground distribution with controlled crosstalk.

More subtle benefits accrue from IMPS. For a given characteristic impedance line, dielectric thickness is less for IMPS than for microstrip or stripline. With equivalent design rules and via size, this improves via aspect ratio and manufacturability.

In plated-up processes, the regularity of the IMPS pattern insures more uniform plating with predictable current density from one design to the next. Substrate bow is reduced because of fewer, thinner dielectric layers, and the stress contributed by solid metal planes is eliminated.

In the case of an IMPS structure fabricated on either side of a dielectric layer, the possibility of test and repair of a complete interconnect structure exists.

In the specific case of flex, and the flex design rules previously described, the topology shown in Figure 2 can be modified to account for the relatively large via catch pads required. This will be discussed later.

12.2.3 Electrical Characteristics

IMPS provides controlled impedance signal transmission with very low crosstalk, because of its largely coplanar transmission line environment with AC ground conductors between adjacent signal conductors. A 50 Ω impedance is easily achieved with reasonable geometries. The exact configuration depends on materials and patterning limitations. A recent four metal layer test vehicle exhibited a 2X - 3X reduction in crosstalk compared with a buried stripline structure of equivalent interconnect density. Compared with the uncontrolled signal line structure of typical flex, the difference would be considerable.

IMPS also provides low impedance power distribution. Decoupling capacitance is supplied by surface mount chip capacitors, such as the AVX low inductance capacitor arrays (LICA) that are effective to several hundred megahertz. Though the parasitic resistance and inductance of IMPS planes is higher than solid planes, the power distribution impedance limit is determined by the size and number of decoupling capacitors, just as in other MCM technologies. This approach is far more cost-effective for MCM-D, for example, than providing separate power and ground planes with an intervening thin dielectric layer. In double-sided flex, IMPS provides electrical characteristics impossible otherwise.

The IMPS structure has been simulated, and results compared with measurement, with excellent agreement. Both the transmission line and power distribution properties are well understood.

12.2.4 Designing in IMPS

A custom addition to the Mentor Graphics MCM Station makes IMPS implementations reasonably transparent to the designer. This system starts with a "phantom" layout of power distribution traces, customized to the particular manufacturing technology. Components are placed and I/O pads defined, usually on a grid compatible with the power distribution grid. The standard MCM router is run to provide signal wiring. Finally, a post-processor is run which provides the necessary adjustments to and clearances around the Boolean combination of signal and power wiring.

2.5 Examples of IMPS

Several different circuits have been implemented in IMPS, particularly in MCM-D technology. In MCM-D the extremely dense wiring capability of 20 μm line and space and 25 μm vias provides high signal line density, even though the signal lines are at twice the minimum design rule (half of the possible density.) One small section of an MCM-D IMPS substrate is shown in Figure 3. The three pads are wirebond pads, and their signal lines are shown tucked in between narrowed mesh conductors.

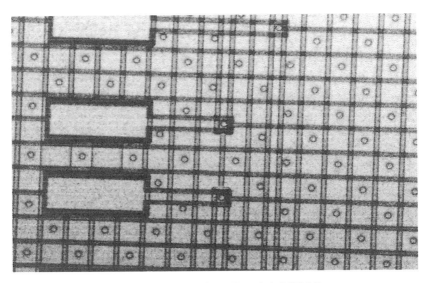

Figure 3. Close up of IMPS mesh in MCM-D.

A wider view of an IMPS MCM-D is shown in Figure 4. In this photo there are not many wide mesh conductors, as the signal line density is extremely high. It is particularly difficult to carry mesh connectivity across a line of wirebond pads, but it can be done, as shown.

Figure 4. Wide view of IMPS mesh in MCM-D.

12.3 IMPS IN FLEX

Fabrication of IMPS in flex has been in the very restrictive environment of Sheldahl's ViaGrid process. In this process, predefined via holes on a, for example, 10 mil pitch matrix, were laser drilled in the flex before any metalization was applied. An example of the design rules that could be used is shown in Figure 5.

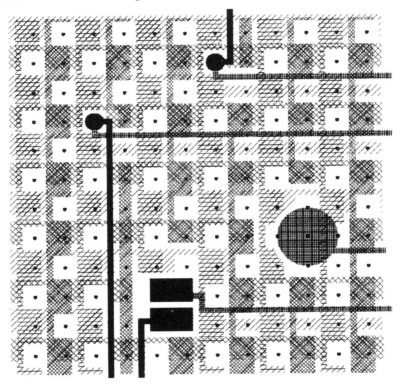

Figure 5. Possible ViaGrid design rules.

The difficulty of designing with ViaGrid is that all unused vias have to be avoided, or even completely eliminated by etching away the undesired copper on both sides of the flex. This prevents shorts between the power and ground mesh structures. The proposed design rules were:

- 250 μm Viagrid pitch
- 50 μm signal lines
- 50 μm spaces

- 250 μm min signal line pitch
- 200 μm P/G lines
- 150 μm round via capture pads
- 200 μm square via keepouts
- Keepouts on both layers
- 500 μm BGA pads
- 100 μm P/G one side of signal
- 200 x 350 μm wirebond pads

In actuality, these design rules were beyond the capability of Sheldahl's process at the time. To realize a real circuit, ViaGrid on 500 μm pitch was used. A photograph of a prototype substrate is shown in Figure 6. Unfortunately, it is impossible to see both sides of the substrate except in a very few areas, because of the high metal coverage.

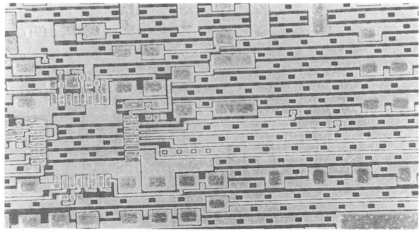

Figure 6. Sheldahl IMPS on ViaGrid implementation.

Wide power and ground mesh conductors have rectangular cutouts 250 μm X 200 μm where every via of opposite polarity would intersect them. Mesh conductors are 300 μm wide, so that a 100 μm wide signal conductor and two 50 μm spaces can fit between them. The cutouts are rectangular so that the connectivity of the trace is maintained by two 50 μm wide sections in the 300 μm conductor width. Signal vias occupy a via site that would otherwise be used for a power or ground via. An isolated catch pad, 150 μm X 200 μm, is used for the signal. Vias are 50 μm in diameter.

This particular design was for an RF/analog circuit with many passive components. Many gold plated solder pads for these components are visible in Figure 6. At the left side is a site for a wirebonded bare IC with 32 leads, 8 per side. Another photograph of a slightly different section of the same design, but assembled, is shown in Figure 7. Following tight assembly design rules can create an extremely dense assembly.

Figure 7. Assembled Sheldahl IMPS substrate.

Several of the signal lines were measured using a Tektronix IPA 510 with TDR function. The impedance was a reasonably consistent 50 Ω, with the change in direction from X to Y invisible because of the consistent geometry of the IMPS coplanar line between directions.

This design was highly constrained because of the predefined via locations that had to be avoided to prevent power distribution shorts. Other technologies are far more flexible and amenable to IMPS. The main problem in any IMPS design is accommodating the vias needed to tie the distribution meshes together. Technologies that require huge catch pads compared to line width do not work well. An example of this problem is shown in Figure 8. In this technology, lines and spaces of 50 μm are no problem, but a 250 μm via catch pad is needed. To get any reasonable line density, it was necessary to run two signal lines between each pair of mesh conductors, with vias on 400 μm pitch. A great deal of conductor "weaving" was necessary to do even that. But

298

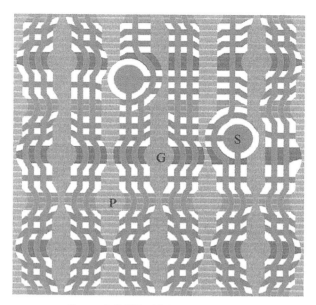

Figure 8. IMPS with large catch pads.

the resulting net potential signal line density, 250 in/in^2, is not terrible, despite the pads, because two lines are run in each track.

Using the design rules for flex stated earlier, namely 50 μm line and space and 150 μm catch pads on 250 μm pitch, it is possible to have the construction shown in Figure 9. Each signal trace again has its own channel between power distribution conductors, and only a slight amount of "weaving"

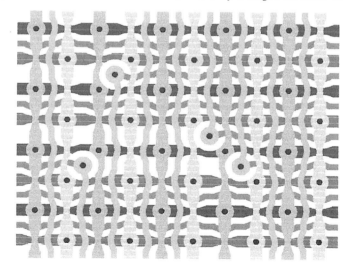

Figure 9. IMPS with flex design rules.

is required. The signal line density is a respectable 200 in/in^2, since there is one signal conductor for every 250 μm of linear dimension.

Thus the IMPS/Flex combination appears particularly well suited for analog and mixed signal or RF applications with moderate signal line density.

12.4 DETAILED ELECTRICAL EVALUATION

The most detailed electrical evaluation of an IMPS structure was carried out on a set of cofired ceramic test vehicles designed at the University of Arkansas' High Density Electronics Center, and fabricated by Kyocera Corporation. The object of the exercise was to compare the performance of two four-metal-layer structures, one fabricated according to the conventional buried stripline topology with Ground, X-signal, Y-signal, and Power layers; the other fabricated as four layers of IMPS. This comparison was chosen because both constructions have the same signal line density. It was expected that the electrical performance of IMPS, especially in crosstalk, might be superior.

12.4.1 Test Vehicles

To perform a fair comparison between IMPS and buried stripline topologies, two four-metal-layer test vehicles were built in cofired ceramic and populated with high speed clock driver chips to simulate simultaneous switching drivers on a VLSI IC. Significant lengths of parallel bus lines were included in each substrate to ensure measurable crosstalk. A photograph of an assembled test vehicle is shown in Figure 10. The stripline version (STRIP) uses two solid metal planes on the bottom (M5) and top (M2) metal layers to distribute power. (Metal M1 is a pad layer.) The IMPS version distributes power throughout the design using all four metal layers.

Both substrates use six IDT 49FCT805 Buffer/Clock Driver dice to drive loads at the end of lines of varying lengths. Each IDT die contains 10 buffers that can be switched on or off in sets of five. This switching capability is exploited so that adjacent lines can be driven or set to a static state to allow accurate measurements of parameters such as crosstalk, ground-bounce, etc. At each end of a line used to connect a driver to a load, there is a 5-point Cascade probe site to allow accurate measurements to be taken.

The line pattern used in the design can be seen in Figure 11. In the STRIP version, all of the lines are on the same metal layer; which layer depends on whether they are oriented in the X or Y direction. In the IMPS version, the lines would alternate between M2/M4 and M3/M5, depending on the orientation of the lines. This configuration allows for multiple test scenarios; for instance, the victim line could be held in a logic-high state while the Load C and Load B lines were being driven by a given signal; probes can then be used to measure the crosstalk seen on the victim line.

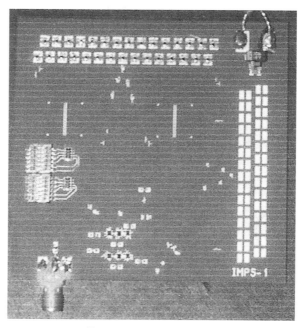

Figure 10. IMPS substrate.

Figure 12 shows a simplified schematic for the design. All non-victim lines are driven by a common input signal; this is the signal that will be used to

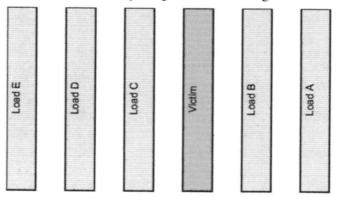

Figure 11. Line for test substrates.

induce noise on the victim lines. Each of the lines being driven by this signal can be enabled/disabled via a DIP switch setting; this allows maximum flexibility for configuration. The victim lines are held at a static level (logic "0" or logic "1", depending on a DIP switch setting), thus allowing the user to measure the amount of noise induced on them.

The components can be seen in Figure 10. The signal input SMA connector and input terminating resistor are at the lower left. Flip chip IDT

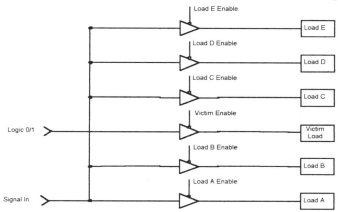

Figure 12. A simplified substrate schematic.

driver chips are surrounded by surface mount decoupling capacitors. DIP switches and pull-up resistors are at the left. Loads (500 ohm resistors) are at the top and right. (The right side is not populated on this test vehicle.) Power leads and bulk decoupling capacitors are at the upper right.

Various amounts of noise and ground bounce can be induced in the power distribution system by setting the DIPswitches to Load Banks A-E. Probe points are provided to measure power distribution disturbances. Crosstalk measurements can be made with any or all of Load Banks A-E switching.

By probing the substrates before components are mounted, using Cascade probes connected to the Tektronix IPA510 TDR, the line impedance and propagation delay can be measured. Crosstalk measurements can also be made on isolated STRIP structures designed to examine the effects of line spacing and the inclusion of interstitial grounds on signal layers. On IMPS, lines designed to examine the effects of multiple vias have been included.

All traces and mesh conductors are 4 mil (100 microns) wide. All traces have, at minimum, a 6 mil (150 microns) space between adjacent traces.

In the IMPS design, the mesh pitch (the pitch between two adjacent power carrying conductors on the same metal layer) is 20 mils.

Kyocera fabricated the substrates, and modules were assembled at HiDEC. Assembly turned out to be extremely difficult because of the tight pitch (125 μm) between I/O pads on the IDT die. The stud-bump bonding method, with gold bumps and conductive epoxy attach, was used.

4.2 Results

Extensive testing has been performed on the test modules as described in the following report sections.

4.2.1 Passive Tests - Power distribution impedance of bare substrates

Power distribution impedance of unpopulated substrates was measured using an HP 4291A impedance analyzer with Cascade ACP40 GSG150 probe (probe points on 150 μm pitch.) A set of probe pads connected to the power distribution system is located near the center of each substrate. Each set of probe pads has two measurement points. The substrate is modeled as a capacitor, using an RLC series equivalent circuit. The model values are given in the table.

	R (mohm)	L (nH)	C (nF)
STRIP-TOP	107	1.6	.65
STRIP-BOT	107	1.6	.65
IMPS-TOP	194	1.5	2.6
IMPS-BOT	139	1.0	2.8

The important feature is that the IMPS substrate has four times as much integral decoupling capacitance as the STRIP substrate. This is caused by the large separation between power and ground planes in the STRIP version, compared with the tight weaving of mesh lines in IMPS.

The differences in R and L between top and bottom IMPS connections are caused by differences in the effective length of wire connecting the pads into the mesh. This is a combination of surface metal (M1) and internal connections. (These values are not the planar R and L of the power distribution meshes.) As has been discovered during this program, it is most important to use as many vias as possible around connection points between pads and mesh to ensure solid, low impedance connections. The IMPS substrate could have been improved with better connections. Both STRIP test points were identical, as was expected.

4.2.2 Impedance and delay of transmission lines

Both IMPS and STRIP produce controlled impedance transmission lines with impedances dependent on the line and ground conductor geometry. Because the STRIP sections were a bit narrow, impedance ranged from 58 ohm to 63 ohm. The IMPS lines had an average impedance of 52 ohm, as measured on the TDR. To avoid impedance discontinuities, it is important to have good coupling from the ground conductor pad into the mesh ground. Both types of line had a propagation delay of 110 ps/cm, as determined by the effective dielectric constant of the material.

4.2.3 TDR crosstalk on striplines at various spacing

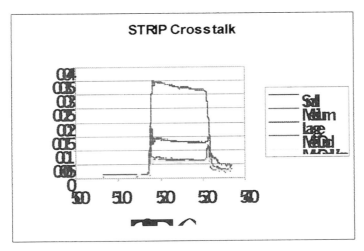

Figure 13. STRIP crosstalk with line spacing.

Five stripline structures were examined for crosstalk using TDR. Each was 6.4 cm in length. Lines at 1X (S), 2X (M), and 3X (L) minimum pitch were measured, as well as lines at 2X minimum pitch with a center ground conductor, with (MGV) and without (MG) intermediate vias. Near end crosstalk for these structures is shown in Figure 13. The results are as expected. An interstitial ground reduces crosstalk between conductors on 20 mil pitch to the level of that between conductors at 30 mil pitch without ground, but with half the signal line density of 10 mil pitch.

4.2.4 Impedance of IMPS lines with various via patterns

As the IMPS transmission line is basically coplanar, it is important to see the effect of a transition from an X-going to a Y-going layer. The impedance of IMPS lines with one X-Y transition in the middle is shown in

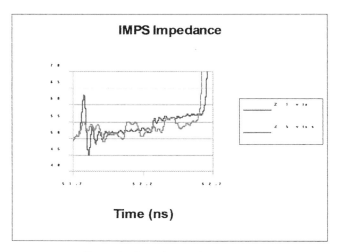

Figure 14. Impedance of IMPS lines.

Figure 14. The mismatch at the driven end is clear and is caused by a short high impedance section of M1 on the surface of the part. The left portion of the trace is in M3, and the right portion in M2. The initial impedance is ~ 50 ohm. The difference between traces on the two layers is only ~ 3 ohm. The difference is caused by the outer layer trace (M2) "seeing" a slightly different electromagnetic environment than the inner layer trace (M3), whose impedance is lower.

The impedance of an IMPS line with five vias (other than vias at the ends) and thus six sections is also shown. In this case, the M1 line section was very short, so no initial impedance mismatch can be seen. The six line sections are clearly visible, with the first section in M2, and the following section in M3, with the sections alternating thereafter. The M2 impedance is ~ 53 ohm, and the M3 impedance is ~ 50 ohm. Though this impedance difference is certainly tolerable, a slight stencil compensation to widen signal traces on the outer layers would match the impedances even more closely. Note that the Tektronix IPA510 used to generate these impedance profiles has a rise time of 35 ps. Most of these variations would be invisible to waveforms with risetimes of 100 - 200 ps.

12.4.2.5 Active Tests - Crosstalk to victim lines with various banks of driven lines

Each substrate has two groups of 30 lines each, the top group and the side group. The coupled length for the top group is longer than for the side group. For this study the top group was used. There are five victim lines, and five banks (A-E) of five loads in each bank, which can be switched to induce varying amounts of crosstalk. The victim lines are enabled and held in a low state.

For all of these measurements, the signal generator was set to deliver a 40 ns period pulse train with 50% duty cycle, with a low voltage of .2 volts and a high voltage of 3.8 volts. This period was chosen so that any ringing on each pulse had a chance to die out before the next transition. As crosstalk is proportional to waveform rise time, the crosstalk peak-to-peak voltage

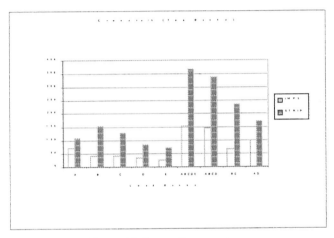

Figure 15. Crosstalk comparison.

depends on the switching time of the 805 drivers, not on the signal generator period. Varying the generator period from 50 ns to 10 ns had almost no effect on peak-to-peak crosstalk voltage.

Top group crosstalk for the substrates is compared in Figure 15. The peak-to-peak crosstalk voltage is indicated for various combinations of banks switching. In all cases the crosstalk for the IMPS version was from 2X to 3X better than STRIP.

12.4.2.6 Power distribution noise with various banks of driven lines

Power distribution noise was measured at the same test point used to measure power distribution impedance. For each substrate, banks A side, A top, B side, B top, etc., were turned on in sequence, and the peak-to-peak noise disturbance measured. Figure 16 shows the power noise, plotted against the number of banks (1 - 10) switched on, for each substrate. No adjustment has been made for the few non-functional output lines.

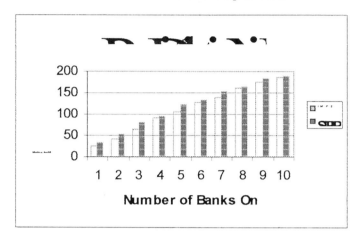

Figure 16. Power distribution noise comparison.

The trend is clear. Noise initially increases proportionally to the number of banks on, and then gradually begins to roll off, as the amount of decoupling on the substrate becomes inadequate to supply the switching currents at speed. (With the 5X attenuators being used on the sampling head, the peak-to-peak noise voltage is around a volt.)

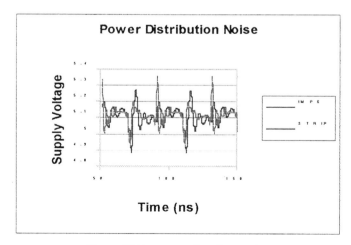

Figure 17. Power supply voltage noise.

In all cases, IMPS noise is lower than STRIP. Figure 17 shows the waveforms of the power noise on the two substrates with three load banks activated. The waveforms were quite similar for any number of banks, only differing in amplitude; three banks is representative.

Comparing IMPS and STRIP, it is clear that the higher integral decoupling capacitance of IMPS has reduced the high frequency peaks of the noise waveform, resulting in the modest reduction in peak-to-peak noise in IMPS. The low frequency noise, whose attenuation is determined by the amount of discrete decoupling capacitance present on the module, is about the same in both cases, indicating that the less-than-optimal connections from capacitor pads into the IMPS mesh did not introduce enough inductance to damage the effectiveness of the discrete capacitors.

12.4.3 A real microprocessor package

Following the evaluation described above, Kyocera engineers redesigned and built an existing microprocessor package using IMPS. The original package had 7 ceramic layers, and required 4 power distribution planes and 2 signal fanout layers. The IMPS package needed only 5 ceramic layers, with 1 ground plane and 3 IMPS layers. An extensive comparative evaluation was conducted. The average signal line length increased with IMPS, from 8.0 mm to 9.2 mm, because of the need for orthogonal rather than radial signal fanout conductors. However the lines maintained good transmission line characteristics, as measured by through pulse transmission.

Crosstalk measurements between adjacent lines were made. As with the active test vehicles, pulse measurements with a 200 ps rise time showed half the crosstalk (16 mV) with IMPS compared with the conventional configuration (31 mV). Measurement of the power distribution impedance showed a small difference; the conventional package looked like a 1.4 nF capacitor, while the IMPS package read 1.1 nF. The conventional package had full Vdd and Vss planes separated by only a single ceramic layer, which explains why its capacitance was higher than the IMPS version.

12.4.4 Conclusions from the electrical test program

The overall conclusion that can be drawn from this program is that substrates using the IMPS topology are at least equal to, and in some characteristics superior to, substrates using a conventional stripline topology. The direct comparison using four metal layer test vehicles is valid because each topology provides the same functionality: low impedance power distribution planes and a total available signal line density of 200 in/in^2. Specific conclusions for the test vehicles include those below.

IMPS supplies 4X more integral decoupling capacitance than STRIP in the four metal layer configuration because of closer coupling between power and ground conductors. This is clear from the passive impedance measurements and from the reduction in high frequency power distribution noise in IMPS.

IMPS lines can be constructed at 50 ohm impedance using standard cofired ceramic design rules, with thinner ceramic than STRIP. The variation in impedance among IMPS layers is small. Impedance discontinuities when going through vias are also small, and manageable with careful design.

Both IMPS and STRIP exhibit the same signal transmission delay.

Crosstalk on STRIP transmission lines can be reduced by spreading lines apart or inserting ground lines between signals, but at the cost of reduced wiring density.

Crosstalk on coupled lines is 2X to 3X lower with IMPS than with STRIP. This is confirmed by multiple measurements with varying numbers of outputs switching.

Connections from bonding or component pads into the IMPS power distribution mesh must be made with care, to avoid introducing inductive parasitics.

For the real microprocessor package, we conclude that an equivalent functional package can be achieved with fewer layers using IMPS technology compared with a conventional design. IMPS moreover gives the advantage of half the crosstalk of the conventional design. Thus the IMPS topology is attractive for the construction of real packages both for performance and cost reasons.

12.5 OVERALL CONCLUSIONS: IMPS AND FLEX

Most of the present applications of foldable flex have modest power distribution, density and crosstalk requirements. Using IMPS technology with foldable flex stretches the horizons to include more system-in-a-package applications for situations that require better crosstalk control and power management.

IMPS has the capability of turning flex into a high-performance interconnect medium. If production technology allows the geometries for signal lines and catch pads postulated, IMPS can provide 200 in/in^2 signal line density with low crosstalk, and low impedance planar power distribution, all on one sheet of double-sided flex. Many low-cost systems, particularly mixed-signal and wireless systems, can be accommodated with that capability. Using foldable flex and IMPS means highly compact systems can be produced without compromising electrical performance.

12.6 IMPS BIBLIOGRAPHY

1.   L. Schaper, S. Ang, Y. Low and D. Oldham, "The Electrical Performance of the Interconnected Mesh Power System (IMPS) MCM Topology," *IEEE Trans. Comp., Packaging., and Manufacturing Tech.*, vol. 18, no. 1, pp. 99-105, February 1995.

2.   M. D. Glover, J. P. Parkerson, and L. W. Schaper, "A Signal-noise Comparison of The Interconnected Mesh Power System (IMPS) with a Standard Four-layer MCM Topology," *Proc. 5th International Conference & Exhibition on Multichip Modules*, pp. 216 - 219, April 1996.

3.   Yee L. Low, Leonard W. Schaper, and Simon S. Ang, "Modeling and Experimental Verification of the Interconnected Mesh Power System (IMPS) MCM Topology", *IEEE Trans. Comp., Packaging., and Manufacturing Tech.*- Part B, Vol. 20, No. 1, February, 1997, pp. 42-49.

4.   J. P. Parkerson and L. W. Schaper, "A Design Methodology for the Interconnected Mesh Power System (IMPS) MCM Technology, Proceedings of the 1995 IEPS Conference, pp. 192 - 196, September, 1995.

5.   Jeff Klein and Armagan Akar, "Multichip Module Offers Ceramic Power", *Electronic Engineering Times*, June 24, 1996, p. 52.

6.   Leonard W. Schaper, "Applications of the Interconnected Mesh Power System (IMPS) Substrate Layer Reduction Topology," in Electronic Packaging for High Reliability; Low Cost Electronics, ed. Tummala et al, Kulwer, 1999.

7.   Leonard W. Schaper, Michael D. Glover, and Masanao Kabumoto, "Comparison of the Interconnected Mesh Power System (IMPS) and Buried Stripline Interconnect Topologies," Proceedings of the International Conference and Exhibition on High Density Interconnect and Systems Packaging, April 2001, pp. 58 – 63.

Chapter 13

Thermal Management and Control of Electromagnetic Emissions

Carl Zweben, Ph. D.
Consultant
Devon, Pennsylvania, USA

13.1 INTRODUCTION

Thermal management has become one of the key electronic packaging design issues. The complex geometries associated with 3D packaging concepts, such as those using folded flexible films, raise new challenges. As microprocessor clock frequencies have increased, electromagnetic emissions have become another important consideration. Typical aluminum and copper finned heat sinks often serve as efficient antennas, exacerbating the problem. The advanced materials discussed in this chapter provide the packaging engineer with efficient ways of addressing both of these issues.

Property improvements of advanced materials include:

- Thermal conductivities ranging from extremely high (over four times that of copper) to extremely low
- Coefficients of thermal expansion that are tailorable from –2 to +60 ppm/K
- Electrical resistivities ranging from low to very high
- Extremely high strengths and stiffnesses
- Low densities
- Low cost, net shape fabrication processes
  These property improvements provide a number of important benefits, including:
- Reduced thickness
- Improved thermal performance
- Reduced thermal stresses and warpage
- Simplified thermal design
- Possible elimination of heat pipes
- Improved fiber alignment in photonic packages
- Weight savings up to 90%

- Improved performance
- Increased reliability
- Reduced electromagnetic radiation emissions
- Increased manufacturing yield
- Potential cost reductions

There are varieties of active and passive approaches that are being used to address the thermal management problem. Active methods include heat pipes, droplet cooling, forced liquid cooling, etc. In this chapter, we concentrate on use of thermally conductive advanced materials, a passive approach, which has the advantages of simplicity, high reliability, absence of power requirements, and, in all probability, lowest volume, weight and cost. For very high power levels, however, active methods may be required.

A key reason for using 3D packaging and thinned chips is to reduce package thickness. Use of advanced materials with high thermal conductivities will result in the thinnest heat sinks. Another benefit is that these materials typically have low densities, so they are also most likely the lightest solution. The weight of a component is obviously important because of its direct contribution to system weight. However, another important consideration is that shock loads arising in shipping and service are directly proportional to mass.

An additional benefit of some of the advanced materials is that they have tailorable coefficients of thermal expansion (CTEs), which can be used to reduce thermal stresses and warping.

Traditional microelectronic packaging materials do not fully meet the needs of many new systems, requiring significant compromises. To address these issues, a significant and increasing number of advanced monolithic and composite materials have been, and are continuing to be developed (1-3). Several are being used in a significant number of relatively high volume commercial systems, such as servers, notebook computers, cellular telephone base stations and power supplies for trains, hybrid automobiles and wind turbine generators. Although the focus of this chapter is on commercial applications, we note that some of the advanced materials also play an important role in aerospace and defense systems.

In this chapter, we discuss traditional and advanced materials, and explore how they can be used in folded flex designs.

13.2 TRADITIONAL PACKAGING MATERIALS

Table 1 presents properties of semiconductors and traditional packaging materials including thermal conductivity, CTE, modulus and specific gravity. It also includes specific thermal conductivity, which we define as thermal conductivity divided by specific gravity. This is a useful figure of merit for applications for which both high thermal conductivity and low weight are important. We note that there are often considerable differences in material properties reported in the literature.

For most applications, packaging materials with high thermal conductivities are required. To assure alignment and minimize thermal stresses and warpage, it is desirable to have packaging materials that match the CTE of semiconductors, ceramic substrates and optical fibers to which they are attached. The CTEs of these materials fall in the range of about 0.5 to 7 ppm/K. Aluminum and copper have relatively high thermal conductivities, but CTEs that are much greater than desired. When these materials are joined to semiconductors and ceramics, relatively thick compliant adhesives are frequently used to minimize thermal stresses. The penalty is increased thermal resistance. Concerns about thermal stresses arising from CTE mismatch often rule out use of solder and braze attachment, which are desirable in some designs because of their thermal conductivities are much greater than those of polymeric materials, even ones having thermally conductive fillers.

The traditional low-CTE packaging materials, such as "Invar", "Kovar", copper/tungsten and copper/molybdenum, have high densities, and are limited to a maximum thermal conductivity of about 200 W/mK. However, advanced materials having thermal conductivities as high as 1500-1700 W/mK, about four times that of copper, and low densities are now commercially available.

The list of traditional metallic materials with CTEs in the range of optical fibers is very short, "Invar" being the leading candidate. This alloy is relatively dense, hard to machine, and has a very low thermal conductivity. On the other hand, several advanced composites have tailorable CTEs that can match that of silica, high thermal conductivities and, for most practical ranges of fiber content, lower densities than that of "Invar". Composites with tailorable CTEs and low thermal conductivities are also available.

13.3 ADVANCED PACKAGING MATERIALS

Increasing a critical parameter by an order of magnitude often has a revolutionary impact on a technology (Consider the effect on transportation of automobile speed compared to that of the horse). Advanced packaging materials offer order-of-magnitude improvements in specific thermal conductivity over traditional low-CTE packaging materials.

There are numerous advanced packaging materials at various stages of development. They fall into six main categories, monolithic carbonaceous materials, metal matrix composites (MMCs), polymer matrix composites (PMCs), carbon/carbon composites (CCCs) and advanced metallic alloys. Some of the latter can be considered metal matrix composites, as discussed later. Tables 2, 3 and 4 present properties of what we believe to be the most important advanced materials at this time. However, we emphasize that we

314

Reinforcement	Matrix	Thermal Cond. (W/mK)	CTE (ppm/K)	Modulus (GPa)	Specific Gravity	Specific Thermal Cond. (W/mK)
Aramid	Epoxy	0.9	1.4	11	1.38	0.6
"Invar"	Silver	153	6.5	110	8.8	17
Cont. Carb. Fib.	Polymer	330	-1.1	186	1.8	183
Cont. Carb. Fib.	Copper	400-420	0.5-16	158*	5.3-8.2	49-79
Cont. Carb. Fib.	Aluminum	218-290	-1-+16	131*	2.3-2.6	84-126
Disc. Carb. Fib.	Aluminum	185	6.0	14	2.5	74
Disc. Carb. Fib.	Polymer	20-330	4-7	30-140	1.6-1.8	12-183
Cont. Carb. Fib.	Carbon	400	-1.0	255	1.9	210
SiC Part.	Aluminum	170-220	6.2-16.2	106-265	3.0	57-73
Disc. Carb.-Graph.	Aluminum	400-600	4.5-5.0	90-100	2.3	174-260
Diamond Part.	Aluminum	550-600	7.0-7.5	-	3.1	177-194
Diamond Part.	Copper	420	5.8	-	5.9	71
Beryllia Part.	Beryllium	240	6.1	330	2.6	92

Table 3. Properties of selected advanced composite packaging materials.

Reinforcement	Matrix	Thermal Cond. (W/mK)	CTE (ppm/K)	Modulus (GPa)	Specific Gravity	Specific Thermal Cond. (W/mK)
"Invar"	Silver	153	6.5	110	8.8	17
Beryllium	Aluminum	210	13.9	179	2.1	100
Silicon	Aluminum	126-160	6.5-13.5	100-130	2.5-2.6	49-63

Table 4. Properties of advanced alloys/metal matrix composites

Reinforcement	Matrix	Thermal Cond. (W/mK)	CTE (ppm/K)	Modulus (GPa)	Specific Gravity	Specific Thermal Cond. (W/mK)
-	Silicon	150	4.1	-	2.3	65
-	Silica Fiber	-	0.6-0.8	72	2.2	0.6-0.8
-	Alumina	20	6.7	380	3.9	5.1
-	Aluminum	218	23	69	2.7	81
-	Copper	400	17	117	8.9	45
-	Epoxy	1.7	54	3	1.2	1.4
-	"Invar"	11	1.3	150	8.1	1.4
-	"Kovar"	17	5.9	131	8.3	2.0
E-glass Fib.	Epoxy	0.16-0.26	11-20	16-19	2.1	0.1
Copper	Tungsten	157-190	5.7-8.3	230-252	15-17	9.1-12.8
Copper	Molybdenum	184-197	7.0-7.1	276-282	9.9-10.0	18-20

Table 1. Properties of silicon, silica fibers and selected traditional packaging materials.

Reinforcement	Matrix	Thermal Cond. (W/mK)	CTE (ppm/K)	Modulus (GPa)	Specific Gravity	Specific Thermal Cond. (W/mK)
-	HOPG	1500	-1.0	-	2.3	650
-	Carbon Foam	100-150	2-3	-	0.2-0.6	250-375
"ThermalGraph"	-	750	-0.5	340	1.76	426

Table 2. Properties of selected advanced carbonaceous packaging materials.

are in the early stages of a very dynamic technology. New materials may well emerge that will eclipse the ones considered here. We note that these materials are at various stages of development and commercialization. As mentioned earlier, some are now being used in an increasing number of relatively high volume commercial and aerospace applications.

Many of the advanced materials of greatest interest are composites. Perhaps the best definition of a composite material was proposed by Kelly; two or more materials bonded together (4).

Composites have been used in electronic packaging for decades. Fiber-reinforced polymer printed circuit boards are one example. In addition, varieties of ceramic and metallic particles are added to polymers to tailor properties such as thermal and electrical conductivity, dielectric properties, CTE, cure shrinkage and viscosity. These are all particulate composites.

In recent years, an increasing number of advanced composites have been developed that offer great improvements in key properties. The most important types of advanced packaging composites consist of polymer, metal and carbon matrices reinforced with fibers, particles, or a combination of the two. The key reinforcements are continuous and discontinuous thermally conductive carbon fibers and thermally conductive ceramic particles, such as silicon carbide and beryllium oxide (beryllia). Many composites, especially those reinforced with fibers, are strongly anisotropic. Particle-reinforced composites are usually, but not always, isotropic. The properties of fiber-reinforced composites in Table 3 are inplane isotropic values.

Carbon fibers merit special consideration. There are dozens of different types. Most have relatively low thermal conductivities. However, there are commercial continuous carbon fibers with nominal thermal conductivities as high as 1100 W/m.K (1-3,5). Experimental discontinuous fibers reportedly have thermal conductivities of 2000 W/m.K. Carbon fibers are being used as reinforcements for polymer, metal and carbon matrices. Because carbon fibers are so thin, it may be possible to bond them to flexible film substrates in the flat state and fold them to the final geometry. We discuss this later.

An important advantage of fiber-reinforced composites is that their properties can be greatly tailored. For example, we can obtain materials with extremely high thermal conductivities in one direction that can compete with heat sinks over short distances. Alternatively, we can achieve high thermal conductivities that are the same in every direction in a plane (planar isotropic) to spread heat effectively. At the same time, we can tailor other properties, such as CTE. We also can obtain materials with thermal conductivities and electrical resistivities that are both much greater than metals. This ability is being used to great advantage in microprocessor heat sinks to reduce electromagnetic radiation.

Most of the advanced materials presented in Tables 2-4 have high thermal conductivities, because of the importance of this property for most packaging applications. However, there are optoelectronic applications for which low thermal conductivities are needed to maintain stable temperatures. There are a number of composites with very low thermal conductivities and tailorable CTEs that meet these requirements. Table 3 presents properties of one example, aramid fiber-reinforced epoxy, a PMC.

Two of the advanced materials in Table 3 are now being used in high volume commercial applications, demonstrating their cost effectiveness; silicon carbide particle-reinforced aluminum (commonly called Al/SiC, AlSiC or Al-SiC), an MMC, and discontinuous carbon fiber-reinforced PMCs. The other advanced materials are at various stages of development and application. However, most are commercially available. A significant advantage of some of the advanced materials is that they can be made by net shape processes, eliminating the need for costly machining.

As mentioned, a particular advantage of composite materials is that their properties can be tailored over a wide range. In one case this was used to solve a warping problem arising in production of a costly ceramic package that caused failure during lid attach, resulting in a scrap rate of over 95% (i.e. the yield was less than 5%). By modeling the complex, multi-step manufacturing processes, we were able to define a CTE range for the composite base plate that limited warping to an acceptable level, increasing yield to well over 99%, resulting in a direct saving of US$ 60 million, and preventing costly litigation.

13.3.1 Monolithic Carbonaceous Materials

Carbon is a remarkable material. It is useful in a variety of radically different forms. Think of diamonds, graphite lubricants and structural carbon fibers used in high performance aircraft and sports equipment. There are several monolithic carbonaceous packaging materials of interest. Table 2 presents the properties of three; highly oriented pyrolytic graphite (HOPG), carbon foam and a rigid fibrous plate called "ThermalGraph".

HOPG, which is also called "thermal pyrolytic graphite" and "annealed pyrolytic graphite", is a highly anisotropic, rather brittle and weak material. However, it has a reported inplane thermal of 1500-1700 W/m.K, about four times that of copper (6). The through-thickness conductivity, however is much lower, about 10-25 W/m.K. The specific gravity of HOPG is only 2.3, compared to 2.7 for aluminum. HOPG has an inplane CTE of −1.0 ppm/K, which is lower than desired for most packaging applications (the negative CTE means that it contracts when heated). However, HOPG can be encapsulated with materials having a variety of CTEs that also provide needed strength and stiffness.

Encapsulated HOPG printed circuit board (PCB) thermal planes (also called thermal cores, heat sinks and cold plates) are being used in spacecraft and aircraft electronic systems.

There is another form of pyrolytic graphite, not considered here, which has a much lower thermal conductivity, but may also be useful for packaging applications.

There are a variety of relatively isotropic carbon foams, which are relatively weak, brittle materials. The most interesting for packaging has cell walls made of carbon having a very high thermal conductivity (7). The result is a highly porous material with a low effective specific gravity (0.2-0.6), an effective thermal conductivity of 100-150 W/m.K and a CTE of 2-3 ppm/K. However, the material probably needs to be protected or strengthened to be of practical use in many applications.

The final type of monolithic carbonaceous material we consider, "ThermalGraph", is composed of oriented discontinuous carbon fibers. Because of the high degree of fiber orientation, its properties are strongly anisotropic. Table 2 presents axial properties.

13.3.2 Metal Matrix Composites

There are a significant number of MMCs developed for electronic packaging (1-3). At this time, the MMC of greatest interest is, by far, silicon carbide particle-reinforced aluminum, commonly called Al/SiC, Al-SiC or AlSiC in the packaging industry. The author was the first to use Al/SiC in microelectronic and photonic packaging, beginning in the early 1980s (8).

Table 3 presents the properties of Al/SiC, which is really a family of materials made by a variety of processes, some of which can produce complex, net shape parts that do not require machining. By appropriate choice of matrix alloy and particle volume fraction, it is possible to tailor the CTE of Al/SiC between 6.2 and 23 ppm/K . Lower values have been achieved with developmental materials. Thermal conductivities range from 170 to 220 W/m.K.

Al/SiC commercial production applications include servers, notebook computers, cellular telephone base stations and power supplies for trains, wind turbine generators and hybrid vehicles such as the Toyota Prius. Production volumes are in the range of millions of piece parts, annually. Al/SiC also is used in a significant number of aerospace and defense applications, such as phased array antennas and printed circuit board heat sinks. The latter include both solid and liquid cooled, flow-through designs.

Table 3 also presents properties of other key MMCs of interest: aluminum and copper reinforced with continuous and discontinuous carbon fibers; aluminum reinforced with discontinuous carbon-graphite and diamond particles; copper reinforced with diamond particles; and beryllium reinforced with beryllia particles.

Carbon fiber-reinforced aluminum thermal planes and microwave packages also have been used in spacecraft electronic systems. Beryllia particle-reinforced beryllium has been used in a limited number of spacecraft applications.

13.3.4 Polymer Matrix Composites

As discussed earlier, glass fiber-reinforced polymers, which are PMCs, have been used in electronic packaging for decades. However, these materials have relatively high CTEs and low thermal conductivities. Their thermal conductivities are so low that they can be considered thermal insulators. In fact, glass fiber-reinforced PMCs are used as thermal insulators in applications such as cryogenic tank supports.

As discussed earlier, a major breakthrough in thermal management technology was the development of thermally conductive carbon fibers. Table 3 presents properties of polymers reinforced with continuous and discontinuous versions (2,3). Continuous fibers produce composites with inplane thermal conductivities 50% greater than that of aluminum, combined with high modulus and low density. PMCs with discontinuous reinforcements tend to have lower thermal conductivities, although there are significant exceptions. This is because discontinuous fibers are inherently less efficient than continuous fibers, and reinforcement volume fractions tend to be lower.

Continuous-fiber PMCs are being used in a number of aircraft and spacecraft production thermal management applications, including electronic enclosures, battery sleeves, radiator panels, and at least experimentally, PCB heat sinks.

A major advantage of PMCs with short discontinuous fibers is that they can be formed into complex parts by injection molding. However, the injection molding process tends to orient fibers, resulting in anisotropic properties. Other processes can produce materials that are approximately inplane isotropic.

There are now a significant and increasing number of production applications of PMCs reinforced with discontinuous thermally conductive carbon fibers, including microprocessor heat sinks, heat spreaders and heat pipe overmolds. Annual production volume reportedly is in the range of millions of piece parts.

An important consideration for PMCs is that polymers absorb moisture, and are therefore not hermetic. In addition, moisture also causes dimensional changes (9). "Popcorning" is an important concern, as is outgassing of organic species.

As discussed earlier, an important advantage of carbon fiber-reinforced PMCs is that they have much higher electrical resistivities than metals, reducing microprocessor heat sink electromagnetic radiation problems.

13.3.5 Carbon/Carbon Composites

Carbon/carbon composites consist of a carbonaceous matrix reinforced with carbon fibers. They are stronger, stiffer and less brittle than monolithic carbon. Some types of CCCs have high thermal conductivities (10). Table 3 presents properties of one formulation.

CCCs have been used in a limited number of production thermal management applications, including launch vehicle PCB thermal planes, spacecraft radiator panels, and thermal doublers for PMC spacecraft radiator panels (11).

13.3.6 Advanced Alloys/Metal Matrix Composites

Table 4 presents properties of three key advanced metallic materials of interest for thermal management, silicon/aluminum, beryllium/aluminum and "Invar"/silver.

As for Al/SiC, silicon/aluminum is a family of materials with a range of properties (12). Since the solubility of silicon in aluminum is low at room temperature, the components separate. Consequently, some consider the materials to be a type of MMC. The situation is similar for beryllium. It is interesting to note that one manufacturer of silicon/aluminum, who called it an alloy for many years, changed the designation to "metal matrix composite" a few years ago.

The silicon/aluminum alloys of greatest interest for packaging are made in billet form by the Osprey process. These materials are reportedly much easier to machine than Al/SiC.

"Invar"/silver is truly a composite material, because it never occurs in an alloyed state during the most common manufacturing process.

13.5   APPROACHES FOR USING ADVANCED MATERIALS IN FOLDED FLEX PACKAGES

As noted earlier, it is well recognized that thermal management is one of the key problems in packaging. This also applies to folded flex designs containing devices that generate significant amounts of heat, such as microprocessors.

As discussed earlier, the focus of this chapter is on passive methods of removing heat from packages using advanced materials in conjunction with convection. The latter can be natural or forced. The key advantages of these materials are high thermal conductivity, low density and tailorable CTE. The primary benefits are reduced thickness, mass, thermal stresses and warping.

Many of the advanced materials discussed in this chapter are electrically conductive, so that some form of electrical insulation may be needed. They share this problem with conventional metallic packaging materials like aluminum and copper. Use of appropriate electrically nonconductive adhesives or thermal interface materials can solve this problem.

There are only three paths to conduct heat from the package; through the top, bottom or laterally to the edges. The latter has been successfully demonstrated by Fraunhofer using copper heat sinks that project out from the package, essentially acting like heat sink fins. These fins allow heat removal by natural or forced convection. Figure 1 demonstrates this concept.

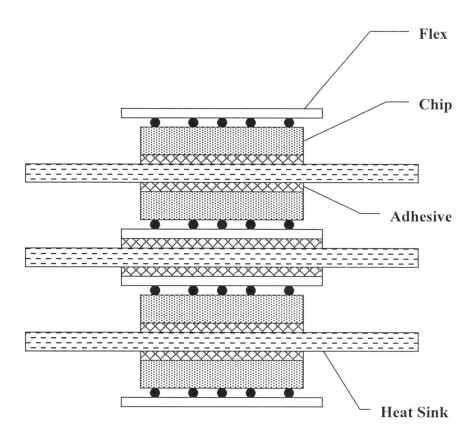

Figure 1. Concept for heat removal using heat sinks.

In this type of design, advanced materials offer the potential for reduce thickness and mass compared to copper. To a first approximation, the thickness reduction is directly proportional to the ratio of inplane thermal conductivities. For example, a HOPG heat sink theoretically would be about 75% thinner than one made of copper. It also would be over 90% lighter, because both the density and thickness are lower. However, very thin HOPG plates may present practical manufacturing problems.

If it is only necessary to conduct heat along one axis, "ThermalGraph" is a key candidate material. Here, again, the material would offer significant thickness and weight advantages over copper and aluminum.

If matching CTE is important, then the composite materials in Table 3 merit special consideration. There are varieties of choices, some of which have higher thermal conductivities than copper, but, at this time, none has conductivities as high as HOPG and "ThermalGraph".

The alternatives to removing heat laterally are to conduct it to the top or bottom of the package. The problem here is that heat must pass through flex substrate materials and adhesives, which are poor thermal conductors. Of course, use of thermal vias can reduce this problem to some extent.

It may be possible to use continuous thermally conductive carbon fibers to help transfer heat to upper and lower surfaces, as well as laterally to the edges. Since the diameter of carbon fibers is very small, it may be possible to bond them to the flex in the flat configuration and bend the film without breaking the fibers. Figure 2 demonstrates this concept.

Since carbon fibers are electrically conductive, insulation must be considered. For one-sided flex, the fibers can be bonded to the circuit-free side. For the two-sided case, an electrically insulating, thermally conductive adhesive is required.

In addition to the possibility of fiber breakage, there are other potential problems that have to be considered, such as film wrinkling and separation. This will require some experimentation. We note that carbon fibers have been used for braided heat straps in place of copper in applications where weight is important. This would be an analogous application.

We are in the early stages of development of both folded flexible film packaging and advanced thermal management packaging materials. If folded flex technology becomes a major force in the industry, it is likely that new thermal management materials will be developed to meet the specific needs of this packaging concept. In addition, methods of incorporating the materials, beyond the approaches discussed here, are likely to emerge.

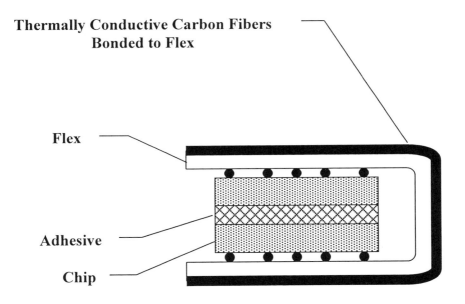

Thermally Conductive Carbon Fibers Bonded to Flex

Flex

Adhesive

Chip

Figure 2 Concept for using carbon fibers to transfer heat vertically and laterally.

13.6 SUMMARY

In this chapter, we examined the increasing number of advanced monolithic and composite packaging materials developed to meet the needs of thermal management and control of electromagnetic emissions. Property improvements over traditional packaging materials, include: extremely high thermal conductivities (up to four times that of copper); tailorable coefficients of thermal expansion (from –2 to +60 ppm/K); extremely high strengths and stiffnesses; low densities; electrical resistivities ranging from low to very high, and low cost, net shape fabrication processes. The payoffs are: reduced thickness; reduced thermal stresses and warpage; lower junction temperatures; simplified thermal design; possible elimination of heat pipes; weight savings up to 90%; increased reliability; improved fiber alignment; increased manufacturing yield and cost reductions. We also discussed ways that these materials can be incorporated into designs based on folded flexible film substrates.

We are in the very early stages of development of advanced composite and monolithic materials that are tailored to meet the specific needs of the packaging industry. If folded flex technology becomes a major force in the industry, it is likely that new thermal management materials will be developed to meet the specific needs of this packaging concept. In addition, methods of incorporating the materials, beyond the approaches discussed here, are likely to emerge.

The history of advanced materials has shown that as production volume grows, costs are reduced, making them increasingly attractive. We have already seen this with Al/SiC in microelectronic packaging. Because of the unique ability of advanced materials, especially composites, to meet future packaging requirements, they increasingly will play a greater role in the 21st century.

13.7 ACKNOWLEDGEMENT

Some of the data in this paper were taken from C. Zweben, "Composite Materials And Mechanical Design", *Mechanical Engineers' Handbook*, Second Edition, Myer Kutz, Ed, John Wiley & Sons, Inc., New York, 1998, and appear courtesy of the publisher.

13.8 REFERENCES

1. D. D. L. Chung and C. Zweben, "Composites for Electronic Packaging and Thermal Management", *Comprehensive Composite Materials*, A. Kelly and C. Zweben, Eds., Volume 6: Design and Applications, Pergamon Press, Oxford, 2000.
2. C. Zweben, "Heat Sink Materials for Electronic Packaging", *Encyclopedia of Materials: Science and Technology*, K. H. J. Buschow, et al., Eds., Pergamon Press, Oxford, In Press.
3. C. Zweben, "Thermal Management and Electronic Packaging Applications", *ASM Handbook*, Volume 21, Composites, ASM International, Materials Park, Ohio, 2001, pp. 1078-1084.

4. Kelly, and C. Zweben, Editors-in-Chief, *Comprehensive Composite Materials*, Six Volumes, Pergamon Press, Oxford, 2000, www.elsevier.com/locate/compcompmat.

5. T. F. Fleming, C. D. Levan, and W. C. Riley, "Applications for Ultra-High Thermal Conductivity Fibers", *Proceedings of the International Electronic Packaging Conference*, Wheaton, IL, International Electronic Packaging Society, 1995, pp. 493-503.

6. M. J. Montesano, "New Material for Thermal Management Has Four Times Thermal Conductivity of Copper", *Mat. Tech.*, Vol. 11, No. 3, 1996, pp. 87-91.

7. J. W. Klett and T. D. Burchell, "High Thermal Conductivity Mesophase Pitch-derived Carbon Foams," 43rd International SAMPE Symposium, May 31 - June 4, Anaheim, CA, 1998.

8. Thaw, J. Zemany and C. Zweben, "Metal Matrix Composites for Microwave Packaging Components", *Electronic Packaging and Production*, August 1987, pp. 27-29.

9. Zweben, "Advanced Composites in Spacecraft and Launch Vehicles", *Launchspace*, June/July 1998, pp. 55-58.

10. W. T, Shih, "Carbon-Carbon (C-C) Composites for thermal Plane Applications", *Proceedings, Seventh International SAMPE Electronics Conference*, June 1994.

11. J. L. Kuhn, S. M. Benner, C. D. Butler and E. A. Silk, "Thermal and Mechanical Performance of a Carbon-Carbon Composite Spacecraft Radiator", *Proceedings, Conference on Composite Materials and Applications*, SPIE International Symposium on Optical Science, Engineering and Instrumentation, Denver, Colorado, July 1999.

12. A. Leatham, "Spray Forming: Alloys, Products, and Markets", *JOM*, Vol. 51, No. 4 April 1999.

Conclusion

John W. Balde, Fellow IMAPS, Fellow IEEE
Interconnection Decision Consulting

This book has presented the state of the technology for thinned silicon chips on thin flexible substrates. The resultant assemblies can be so thin they can be placed on a substrate with metal traces over the chips. They can be placed on strips of flexible tape circuits and tested before folding the flex into the stacked assembly.

These two attributes have made this technology capable of dense but medium complexity circuits for automotive and medical applications. Fraunhofer IZM and Fraunhofer IBMT have exploited the capabilities to make many current products.

This has resulted in an interest in the technology by leading electronic manufacturers, including NOKIA and IMEC. As a result there is a European Project with a team headed by Fraunhofer ISIT to define the processes of thinning and produce standard technology.

But present uses for this technology have been for single chips and stacked memory with single microprocessors on top. We, the authors of this book, think there are additional possibilities.

Evan Davidson makes the point that System-on-a-Chip is a technology that cannot deliver low cost assemblies - that System-in-a-Package has greater capability. But present System - in - a - Package has been made with expensive MCM-C and MCM-D technology. Even MCM-L on single rigid substrates is costly and takes considerable area..

I present the case that the demand for higher and higher speeds and capability has abated – there is a shift to a search for lower cost ways to make the multichip modules needed for System-in-a-Chip. We, the authors of this book, believe that Folded Flex technology with thinned silicon chips can produce much lower cost 2 ½ D technology that is more suitable for mass manufacturing, and has the advantage of not needing known good die.

It is time to consider using the folded flex technology to make multichip packages of greater complexity using mixed chips, not all memory. The greater complexity of multiprocessor chips make very few needed for many applications, often just one. The folded flex constructions are particularly suitable for stacked assemblies of this type.

To do this requires fine line flex at low cost, and the use of LCP technology to enable even lower cost and finer line and space through smaller capture pads around the vias.

Embedded microprocessors need power and ground control which cannot be provided with a second power and ground circuit connected by "Z" axis links, but must have power and ground and crosstalk control in the double sided flex circuit. But that is possible with the IMPS Technology.

Ah, but then heat comes from chips not all on the surface of the stack. Use carbon fibers to dissipate the heat, or embed the assembly in Aluminum / silicon carbide.

All this leads to a possible expanded use of the foldable flex and thinned silicon chip technology – add finer line, add LCP material, add IMPS and add carbon fiber heat paths.

It is our hope and firm belief that this technology has great potential for increased use, to become the major manufacturing technology for System-in-a-Package assemblies.

Author Biographies

John W, (Jack) Balde, Interconnection Decision Consulting, is an IEEE Fellow and an IMAPS Fellow. He has been active as a consultant to over 200 Electronics companies, and active in the Electronic Packaging Societies.

He was one of the early chairmen of the IEEE Computer Packaging Committee, now TC-14 System Packaging Committee, where he now serves as workshop coordinator. A founder of the International Electronic Packaging Society, he served as Chairman of the Board before its merger with ISHM to form IMAPS. He furthered the thrust for surface mount packaging with his focus on chip carrier technology; he nurtured the MCM technology with workshops and conferences, and is currently focusing on foldable flex technology for high capability, low cost manufacture. In his activities for societies he has organized over 40 workshops and 12 conferences - founding the IMAPS MCM conferences in Denver. He has 16 patents in electronic curve tracers, tantalium thin film technology and laser via drilling. He has authored or edited 4 major books and contributes to many others, and presented over 150 technical presentations at major electronics packaging conferences. He is a graduate of Rensselaer Polytechnic Institute, and was for many years an engineer and later a Research Leader for Western Electric company and Bell Telephone Laboratories, before founding his consulting firm, Interconnection Decision Consulting.

Dr. Karlheinz Bock studied electronics and communication engineering at the University of Saarbrücken, Germany. Since January 2001-present he is employed at the Fraunhofer Gesellschaft, Institute for Reliability and Microintegration in Munich, Germany as head of the Polytronic Systems Department working on the development of flexible systems for ubiquitous networking on the basis of thinned flexible semiconductors and polymer electronics on flexible substrates. The department focuses on the development of semiconductor CVD for electro-optical MST and high-frequency electronics, thinning, advanced separation and handling of semiconductor substrates, system assembly, the development of reel to reel production processes in the reel to reel demonstration center and the development of polymer electronic devices and circuits. Applications are i.e. smart clothes, wearable computing, PDA, mobile computers and terminals, smart tags, transponders and labels.

Dr. Bock published or contributed to over 40 scientific papers, 15 invited presentations at conferences, research organizations and companies and 3 books. He holds / has applied for 3 patents in the field of reliability and polymer electronics. He is or has been multiple member of the technical program committees of EOSESD, IRPS, ESD-Forum, GaAs-Mantech and he is a member of several sections of the IEEE

Bill Carlson has been in the electronics industry for his entire career, starting with disk drives at IBM and winding up in semiconductor packaging at Tessera Technologies. Experience includes a number of different roles providing perspectives as a Development Engineer, Program Manager, Business Planning Manager, Marketing Director, VP of Engineering, and CEO of a test equipment company. Education includes MS in Physical Chemistry; technical interests include materials and processes used in electronics. Professional activities include Chairman of ANSI subcommittee X3B7 (magnetic media) and ISO delegate for several years, co-founder and 10-year Board Member of IDEMA trade association, and invited consultant to Department of Commerce Bureau of East/West Trade setting export controls for computer technology. Personal interests involve building and using computers, photography, firearms, maintaining a fleet of 6 family cars, and trying to keep up with his wife's list of home improvement projects.

Evan Davidson recently retired as a Distinguished Engineer from IBM's facility in Poughkeepsie, New York U.S.A. He was a practicing engineer for 40 years. During this period, he did bipolar and CMOS circuit design, chip physical design, electrical package design and high-performance technology applications work. For most of the last 25 years, Evan coordinated the efforts of mainframe system designers and technology designers as a catalyst for choosing optimized product design points. During the '70s and '80s, he was quite involved with defining and designing IBM's ceramic MCM technology at IBM's East Fishkill location. During this time, Evan was influential in pioneering many signal integrity design aid tools. During his career, Mr. Davidson has given scores of presentations, written many articles and book chapters on the subjects of package design and signal integrity. He holds 12 U.S. patents. He is also a Fellow of the IEEE and a member of IMAPS. He is a graduate of Rensselaer Polytechnic Institute and New York University with a BSEE and MSEE, respectively. Evan is also a member of the Eta Kappa Nu and Tau Beta Pi Honorary Engineering Societies.

Michael Feil studied physics at the Technical University of Munich. From 1976 -- 1981, he spent 5 years at Siemens AG. At the beginning, he was involved with the development of electro-mechanical filters for communication applications and later on with the research and development of CCD's for analog signal processing. Since 1981 he is has been with the Fraunhofer Society. From 1981 until 1997 he was leader of the group hybrid microelectronics, assembly and packaging technology at the Institute for Solid State Technology (IFT). Since 1998 he is dealing with the topic of extremely thin ICs for "stand alone"-applications. After termination of the IFT, he became leader of the group of assembly of thin chips & hybrid integration at the Fraunhofer-Institute for Reliability and Microintegration (IZM) Munich division, department Polytronic Systems. In 2002 he is starting the reel-to-reel application center for the development of production principles for thin flexible electronic systems. Michael Feil holds several patents in the field of assembly techniques.

Thomas Harder studied physics at the University of Kiel and received the degree of Diplom-Physiker in 1989. Since 1989 he is working in the field of advanced packaging and interconnection technology. In 1996 he joined the Fraunhofer-Institute for Silicon Technology (ISIT) in Itzehoe near Hamburg heading a group on ´MEMS and MCM Packaging´. He has coordinated a number of multinational European projects including the FLEX-SI project on ultra-thin packaging solutions using thin silicon.

Terry F. Hayden received his Ph.D. in Chemical Engineering from University of Wisconsin, Madison, in 1986. He worked in IBM Microelectronics Development Lab on TAB and flip chip bonding in Austin for 10 years, before working on semiconductor plasma etch and lithography for Cypress Semiconductor in Round Rock, Texas. He has been working on flex circuit packaging for the past 5 years at 3M and is currently in the Microelectronics System Division as a product development specialist. He has authored more than 30 external technical publications and has been awarded five patents. Dr. Hayden can be contacted at 3M Austin Center, 6801 River Place Blvd., Austin, TX, 78726, Tel. 512/984-5682 or e-mail at tmhayden@mmm.com.

Christine Kallmayer received a diploma in experimental physics at the University of Kaiserslautern in 1994. Afterwards she worked as a research scientist at the research center „Technologien der Mikroperipherik" of the Technical University of Berlin. Her main field of activity was the development and investigation of packaging technologies with the Au-Sn metallurgy for different applications, e.g. optoelectronics, chip on flex, chip scale packages and the reliability of the metallurgical system. Since 1998 she is responsible for the group „Flex Circuit Applications" at Fraunhofer IZM. The main working areas are flip chip soldering and adhesive joining on flex, chip size packages and reliability investigations for different contact metallurgies.
During her time at the institute she has presented 10 technical papers as main author at different international conferences. Additionally she co-authored over 20 technical papers.
She is also involved in 7 submitted patents. In 2002 she received the Outstanding Young Engineer Award from IEEE CPMT.

Christof Landesberger, born 1963, studied physics at LMU University in Munich, Germany. He received his diploma degree in 1990 presenting an experimental work on grazing incidence x-ray scattering at single crystalline, ion implanted metal surfaces. The same year he joined the Fraunhofer Society and worked on material analysis, hybrid integration on ceramic substrates and wafer bonding technology. Since 1999 he manages the working group "wafer preparation" at the Munich division of Fraunhofer Institute for Reliability and Microintegration IZM. Research work concentrates on wafer thinning and surface preparation techniques, handling techniques for ultra thin wafers and material characterisation of ultra thin semiconductor substrates. C. Landesberger initiated and organises the annual international workshop on "Thin Semiconductor Devices – Manufacturing and Applications".

Jörg-Uwe Meyer was born in Darmstadt, Germany, in 1956. He received his Engineering Degree from the Fachhochschule Giessen, Germany, in 1981 and his Ph.D. degree in Biomedical Engineering from the University of California, San Diego, USA, in 1988. In 1989, he was awarded a NRC fellowship as a principal investigator at the NASA-Ames Research Center, Moffett Field, USA. He obtained a BBA degree in 1999 from the Graduate School of Business Administration in Zurich. From 1990 -2002, he was heading the Sensor Systems / Microsystems Department of the Fraunhofer Institute for Biomedical Engineering, St. Ingbert, Germany. Beginning August 2002, he is head of research at the biomedical and safety device company Draeger, in Luebeck, Germany. Since 1999, he holds a position as full professor at the University of Saarland. His expertise is on biomedical and industrial sensor developments as well as on telemedical homecare applications and services. His particular research interest comprises biomedical microdevices, neuroprosthetic microimplants and biocompatible hybrid integration and packaging of microcomponents. He holds several patents in the field of microsensors and biomedical instrumentation.

In 1969, **Georges Rochat** obtained a diploma in micro-engineering from the St-Imier Engineering school (Switzerland). In 1971, he obtained his master in micro-engineering from the university of Neuchâtel, Switzerland. In 1974, he created Demhosa, a company specialized in mechanics and sold it in 1999. Then he worked for as a researched engineer at the Swiss watch research laboratory and in 1980 - 1981, he worked as a senior research engineer at SRI International, Menlo Park, California. Then, he became a vice-president of the Swiss Watch research laboratory, before starting Valtronic in 1982. Today, Georges Rochat is President and CEO of the Valtronic group, he also is a member of the Swiss Science Academy and a member of IEEE, as well as a member of the board of a Swiss technical school.

Dr. Leonard W. Schaper is Professor of Electrical Engineering at the University of Arkansas, where he has led a research program at the High Density Electronics Center (HiDEC) in advanced interconnect technologies, including 3-D packaging, advanced heat removal, mesh plane power distribution, MCM-D/L process development, integral passives, ultra-low inductance decoupling capacitors, and through-silicon interconnects. He has been active in electronic packaging since 1980, both at AT&T Bell Laboratories and at Alcoa Electronic Packaging, before joining the university in 1992. Dr. Schaper holds ten patents and has authored or co-authored over 250 talks and papers. He is a Past President of the International Microelectronics and Packaging Society (IMAPS), and is a past member of the IEEE CPMT Board of Governors. Dr. Schaper is a Fellow of the IEEE and a Fellow and Life Member of IMAPS. He is a recipient of the IEEE CPMT Society Outstanding Sustained Technical Contributions Award and the IMAPS William D. Ashman Award.

Ted Tessier is currently the VP of Advanced Applications Development at Amkor Technology located in Chandler Arizona. He is responsible for 3D packaging

336

including stacked die chip scale packaging (S-CSP) and stackable package development as well as MEMS packaging. Over his 18 years in the semiconductor industry, Ted has worked in an assortment of senior engineering and management positions at Nortel / Bell Northern Research, Motorola and most recently Biotronik. Ted has published widely in the areas of advanced packaging including thin film and laminate based substrate technologies, flip chip assembly, medical electronics, wafer level CSP technologies and 3D packaging. He is a Senior Member of the IEEE and IMAPS and is active in the engineering community.

E. Jan Vardaman, President and Founder, TechSearch International, Inc., Austin, Texas. Ms. Vardaman analyzes international developments in the field of semiconductor packaging and assembly. Previously she served on the corporate staff of Microelectronics and Computer Technology Corporation (MCC) in Austin, Texas where she analyzed international developments in software including artificial intelligence and semiconductor packaging and assembly. She is the editor of *Surface Mount Technology: Recent Japanese Developments*, published by IEEE. She is a columnist with *Circuits Assembly* magazine, and author of numerous publications on emerging trends in semiconductor packaging and assembly. She served on the NSF sponsored World Technology Evaluation Center (WTEC) study team involved in investigating electronics manufacturing in Asia. She is a member of IEEE's CPMT society Board of Governors, IMAPS, and SMTA. Ms. Vardaman received her B.A. in Economics and Business from Mercer University in 1979 and her M.A. in Economics from the University of Texas in 1981.

Michael W. Warner is Vice President of Technology and Strategic Programs for Tessera. Since joining Tessera in 1994 as Vice President of Product Development he has been responsible for developing products using Tessera's micro ball grid array semiconductor technology. He has at various times been responsible for Engineering, Manufacturing, Sales and Quality Assurance. Prior to joining Tessera Mr. Warner spent 29 years in the hard disk drive business. He began at IBM where he held various technical, management and executive roles during his 19-year tenure. During the next ten years Mr. Warner served as Vice President of Engineering at disk drive companies Micropolis, Maxtor, Toshiba America and was the co-founder and President of Orca Technology. Mr. Warner holds a B.S. degree in Mechanical Engineering from San Jose State University and holds numerous patents in disk drive technology and semiconductor packaging technology.

Rui Yang received his MS in Physics in 1987 from the University of Missouri, Kansas City and his Ph.D. in materials science and engineering in 1992 from the University of Minnesota, Minneapolis. He joined the 3M Electronic Products Division in 1993 and currently he is a senior research specialist in the 3M Microinterconnect Systems Division. His career has been focused on the electronic packaging industry. He has authored and co-authored more than 40-refereed journal and conference papers, and has been awarded five patents. Dr. Yang can be contacted at 3M Austin Center, 6801 River Place Blvd., Austin, TX, 78726, Tel. 512/984-2530 or e-mail at ryang@mmm.com.

Dr. Carl Zweben, an independent consultant on advanced packaging materials and composites, was for many years Advanced Technology Manager and Division Fellow at GE Astro Space, which was acquired by Lockheed Martin. Other affiliations have included Du Pont, Jet Propulsion Laboratory and the Georgia Institute of Technology NSF Packaging Research Center. Dr. Zweben was the first, and one of only two winners of both the GE One-in-a-Thousand and Engineer of the Year awards. He is a Fellow of ASME, ASM and SAMPE, an Associate Fellow of AIAA, and has been a Distinguished Lecturer for AIAA and ASME. Dr. Zweben began working on aramid printed wiring boards (PWBs) at Du Pont in the 1970s. He continued the development of advanced composite packaging as director of the GE Aerospace Group Advanced Composites Center of Excellence, where he worked on low-expansion PWBs, a variety of advanced thermal management materials, and developed the first silicon carbide particle-reinforced aluminum (Al/SiC) microelectronic and optoelectronic packages. His interests also include application of composites in high speed, precision assembly equipment components. He has to his credit over 100 contributions to journals, handbooks and encyclopedias and has presented over 100 invited lectures, including one at the AIAA 50th Anniversary "Learn from the Masters" series. Dr. Zweben is Co-Editor-in-Chief of the 6-volume, *"Comprehensive Composite Materials"*. He has directed and lectured at over 150 classroom, satellite broadcast and videotape short courses in the US and Europe, including courses on advanced packaging materials for IMAPS, Semi-Therm, NEPCON, SPIE and several companies. Dr. Zweben's clients have included Nokia, ITT Aerospace/Communications, COM DEV, Reynolds Metals Company, Hitco Carbon Composites, Boeing, Hughes, General Dynamics, E-Systems, Brunswick Corporation, Princeton University High Energy Physics Group, US Air Force, Army, Navy and many other organizations.

Index

Index

Index

Index